The Hidden Curriculum—Faculty-Made Tests in Science

Part 2: Upper-Division Courses

Sheila Tobias and
Jacqueline Raphael

California State University, Dominguez Hills
Carson, California

PLENUM PRESS • NEW YORK AND LONDON

Library of Congress Cataloging in Publication Data

Tobias, Sheila.
The hidden curriculum—faculty-made tests in science / Sheila Tobias and Jacqueline Raphael.
    p.   cm.—(Innovations in science education and technology)
Includes bibliographical references and index.
Contents: pt. 2. Upper-division courses.
ISBN 0-306-45581-1
    1. Science—Study and teaching (Higher)—United States—Examinations, questions, etc. 2. Educational tests and measurements—United States. 3. Grading and marking (Students)—United States. 4. Educational change—United States. 5. College students—United States. I. Raphael, Jacqueline. II. Title. III. Series.
Q183.3.A1T58   1997
507′.1′1—dc21   97-1865
                                                                                                                          CIP

ISBN 0-306-45581-1

© 1997 Plenum Press, New York
A Division of Plenum Publishing Corporation
233 Spring Street, New York, N.Y. 10013

http://www.plenum.com

10 9 8 7 6 5 4 3 2 1

Printed in the United States of America

The Hidden Curriculum—
Faculty-Made Tests in Science

Part 2: Upper-Division Courses

INNOVATIONS IN SCIENCE EDUCATION AND TECHNOLOGY

Series Editor:

Karen C. Cohen, Harvard University, Cambridge, Massachusetts

A Continuation Order Plan is available for this series. A continuation order will bring delivery of each new volume immediately upon publication. Volumes are billed only upon actual shipment. For further information please contact the publisher.

PREFACE TO THE SERIES

The mandate to expand and improve science education for the 21st century is global and strong. Implementing these changes, however, is very complicated given that science education is dynamic, continually incorporating new ideas, practices, and procedures. Lacking clear paths for improvement, we can and should learn from the results of all types of science education, traditional as well as experimental. Thus, successful reform of science education requires careful orchestration of a number of factors. Technological developments, organizational issues, and teacher preparation and enhancement, as well as advances in the scientific disciplines themselves, must all be taken into account. The current prospects look bright given national reform movements such as the National Academy of Science's "Standards for Science Education" and the American Association for the Advancement of Science's "Benchmarks"; the backing of science education leadership; and recent developments, including the Internet and new educational software. Further, we have a world-wide citizenry more alert to the need for quality science education for all students, not just those who will become scientists. If we can isolate and combine such factors appropriately, we will have levers for science education reform. The books in this series deal in depth with these factors, these potential levers for science education reform.

In 1992, a multidisciplinary forum was launched for sharing the perspectives and research findings of the widest possible community of people involved in addressing the challenge. All who had something to share regarding impacts on science education were invited to contribute. This forum was the *Journal of Science Education and Technology*. Since the inception of the journal, many articles have highlighted relevant themes and topics: the role and importance of technology, organizational structure, human factors, legislation, philosophical and pedagogical movements, and advances in the scientific disciplines themselves. In addition, approaches to helping teachers learn about and use multimedia materials and the Internet have been reported. This series of volumes will treat in depth consistently recurring topics that can support and sustain the scientific education enterprise and be used to raise levels of scientific knowledge and involvement for all.

The first four volumes illustrate the variety and potential of these factors. *The Hidden Curriculum—Faculty-Made Tests in Science: Part 1, Lower Division Courses* and *The Hidden Curriculum—Faculty-Made Tests in Science: Part 2, Upper Division Courses* are premised on the belief that testing practices influence educational procedures and learning outcomes. Innovations in exam practices that assess scientific understanding in new and more appropriate ways should be shared with the widest possible audience. The research described and the resulting compendium of hundreds of contributed, annotated best exam practices in all science courses at the college level is a resource for every science educator and administrator.

Web-Teaching: A Guide to Designing Interactive Teaching for the World Wide Web aids instructors in developing and using interactive, multimedia educational materials on the World Wide Web. It also helps instructors organize and control these resources for their students' use. Not only do instructors learn how to improve their own materials and delivery, but they can access and make available Web-based information in a way their students can comprehend and master. Using the lever of instructional technology is an increasingly important part of science teaching; this book guides that process.

Finally, *Internet Links for Science Education: Student–Scientist Partnerships* illustrates the workings and effectiveness of this new paradigm and growing force in science education. In these partnerships (SSPs), students help scientists answer questions that could never before be fully addressed due to the lack of a large number of strategically positioned observers. Students gather and analyze data in projects involving authentic and important scientific questions, and science teachers actively explain science to students and help scientists implement their research. Data gathering and sharing, the heart of effective SSPs, is possible and rapid with the help of the Internet and a variety of technologies—groupware, visualization, imaging, and others. Several representative SSPs are described in depth. Chapters on student data and the human and technological infrastructures required to support SSPs help readers understand the interplay of the several factors in this approach to improving science education K–12. The Appendix contains a useful annotated list of current projects with complete contact information. Readers of this book will come away with an understanding of these programs from multiple perspectives and will be encouraged to become involved in similar efforts.

It is our hope that each book in the series will be a resource for those who are part of the science reform effort.

Karen C. Cohen
Cambridge, Massachusetts

PREFACE TO PART 2

Part 2 of *The Hidden Curriculum—Faculty-Made Tests in Science* features innovative testing practices devised and used by faculty teaching upper-division science courses in biology, chemistry, physics, and some engineering and other fields. Some are suprisingly simple and obvious; others are complex, involving the instructors in a reevaluation both of the courses they teach and of their students. In all cases, the effect on their science classrooms, at least according to the progenitors, has been profound.

We can expect upper-division students to be even more surprised by (and perhaps unprepared for) innovative in-class testing, having been exposed, in most cases, to traditional assessments for most of their education. Indeed, many of our contributors speak of a need now to "teach to the test" in an affirmative sense: to educate students to new ways of approaching, preparing for, and taking these altered examinations. From the contributors' point of view, this kind of teaching involves faculty and students in more discussion, more group work, more writing, more research, and a greater familiarity, through reading and discussion, with the research literature. This is one source of the improved exams' impact.

Upon completing the lower-division collection of exam innovations, we asked a few faculty members for comments on the collection. These essays, collectively entitled "From Innovation to Change," are included as the introduction to this collection.

For those unfamiliar with Part 1 of this title, allow us to recapitulate our method for collecting contributions. We began by limiting our study to testing in the natural sciences, except where a submission in mathematics or engineering appeared to the authors to be deserving and applicable to science. A solicitation went out to a variety of mailing lists and through the 1994 Chautauqua Short-Course Catalog to science faculty.[1] In addition, friends and colleagues distributed copies of the solicitation and advertised it in a variety of educational journals.

---

[1] Copy of the solicitation is included in the Appendix.

When soliciting "new theory, new practice," we asked college science educators to describe their innovations in terms of one or more of the following categories:

1. Exam design (including content in general and test item construction specifically).
2. Exam format (such as length, open-ended, closed, multiple choice, verbal, pictorial, quantitative).
3. Exam environment (individual, group, in-class, take-home).
4. Grading practices (retesting, curve grading, resurrection points).

We decided on these categories because in-class examinations are more than the sum of the questions or problem sets faculty invent to measure student mastery. Examinations embed assumptions about student learning, motivation, and attitude, as well as course goals and requirements. Although conscientious faculty generally spend long hours preparing the *content* of their exams, these other matters—exam format, exam environment, and grading practices—get less attention despite the fact that they may be even more salient (in the minds of their students) than the content itself. Building on these categories, we allowed the faculty to describe their innovations in terms of their fundamental parts. Many are made up of a number of ideas that can be modified and employed by faculty in different combinations.

Upon receiving a description of a particular innovation, we standardized the format of the summaries and went back to the innovator, by mail and telephone, for further detail and elaboration. Eventually, each submission included in this book was approved by the originator.

Throughout, we have had the cooperation of our contributors in assisting us in conveying accurately to our readers what they are doing, what they are thinking about what they are doing, what is succeeding, and what is not. We also asked them to detail obstacles they and their colleagues are experiencing while attempting change in in-class examinations at colleges and universities. We are grateful for their enthusiasm for this project and their willingness to share their ideas and thoughts; and to our colleages Herbert Lin, Mary Nakhleh, Susan Nurrenbern, and Arthur Ellis, whose comments comprise the Introduction to this volume. Thanks are also due to Abigail Lipson, of the Bureau of Study Counsel, Harvard University, whose theoretical approach to assessment guided our thinking from the outset. Her written contribution appears as the Commentary at the conclusion of this volume.

Finally, we wish to thank Francis Collea for his support, and the California State University Foundation for its willingness to publish this work.

CONTENTS

CHAPTER 1

INTRODUCTION: FROM INNOVATION TO CHANGE

We asked a number of colleagues to comment on the difficulties involved in moving the exam issue from the realm of innovation to that of systemic change. Following are comments by Herbert Lin, Mary Nakhleh, Susan Nurrenbern, and Arthur Ellis

RETHINKING THE PURPOSE OF EXAMS

Herbert Lin, Physicist and former Physics Educator at MIT

As the first volume of this series demonstrated again and again, institutional needs have a profound effect on the frequency, structure, and format of exams. The multiple-choice exam is easiest to grade and thus makes considerable administrative sense when graders are in short supply. Indeed, many of the faculty writing about their innovations in Part 1 suggest that their innovative exams are unlikely to be adopted by other faculty because of the extra effort and time required as compared to standard exams. Many would concede that even an essay exam, administered to an entire class at the same time, is not as good in assessing student competence as an extended apprenticelike interaction between student and teacher over the course of several months. But apprenticeships in every subject just don't make sense at the undergraduate level, where students are responsible for multiple courses and areas of study. Coupled with faculty concerns over fairness and reliability (not entirely unreasonable in an era of lawsuits filed by students concerned about discrimination of various kinds), it is not surprising that exams evolve slowly, if at all.

Yet for students, as the section "Listening to Students" in Part 1 makes quite clear, exams are fundamental. It is the rare student for whom the exams (and the course grade) are not the central focus of his/her efforts. This is

1

especially true in required courses that students are not intrinsically interested in taking. Since they rely on grades for feedback about how well they have mastered the material, as well as for information about how they compare to others, exams are the quintessential performance. This is why many students "power-cram" before the exam (an approach that no teacher advocates) and throw out their study notes soon afterwards. Another way of making sense of student behavior is that for them, exams are a kind of "marketplace" in which students offer a certain amount of work in exchange for a certain grade.

To illustrate, Arnold Arons tells a story about a testing strategy he used in an introductory physics course to compel students to take seriously the most important concepts in the course. To emphasize the importance of these concepts, he told students they would be tested repeatedly on them throughout the course; that is, a certain topic might reappear on a whole series of exams. To monitor student progress he kept track of how students performed on each question in each exam. When Arons asked a student, who had repeatedly failed the same question, why he had not studied that particular concept, the student replied, "I thought it would go away." He apparently felt that once he had been exposed to certain material on an exam, it was "unfair" to subject him to double jeopardy by testing him again on the same material.

Given that exams will always be an intrinsic part of the undergraduate science course, one approach is to change the exams, as contributors to this and the previous volume are doing; another is to rethink the purposes exams are meant to serve and why they are structured the way they are. This would permit even "old" exams and "old" grading practices to be the basis of new assessment. For example, in a multiple-choice exam designed to test problem-solving ability in physics, a distracter can be a key element of a valid assessment of what a student knows. Such an exam can give points not only for the correct selection, but also for a written explanation of why a certain set of distracters has to be wrong. Of course, grading such essays, like grading problem sets, takes additional effort. But at least the format needn't be altered.

To eliminate internal competition among students being graded on a curve, the late Margaret MacVicar, physics professor at MIT, used former students as tutors in the courses she was teaching. In addition to attending all lectures and assisting current students in the course, they had to take all exams. Grades for current students were based on the average score received by the tutor on the following metric: An A was an exam that received a score equal to 90 percent of tutor average, a B would be 80–90 percent of tutor average, and so on.

Such a grading scheme had a number of advantages. Because grades were tied to an external metric (the tutor average), students did not compete with each other for a limited number of high grades. At the same time, tutor performance on the exams served as a normalizing or calibrating influence on the difficulty of the exams. But perhaps its most lasting benefit was that students, as a group, began to compete with the tutors, often with the result that by the end of the course, their class average exceeded that of the tutors.

Finally, the "baggage" that students bring to our courses presents a formidable challenge to teachers trying to test students in innovative ways. In an independent study course for introductory physics, I told my two students that I would give them an oral exam focused on a specific problem. I told them the problem was too hard for them at their then-current level of expertise, but that in recognition of the independent study nature of the course, I said they should consult with other students, other faculty members, various textbooks, and any other resources that they felt might help until they could talk about a solution to the problem with facility and skill. Two weeks later, I gave them the oral exam and found them hopelessly unprepared. Upon further discussion, I found they had not even talked to each other! When I asked why they had not, when I had specifically asked them to consult resources other than themselves, they said, "Consulting with others would have been cheating," and that cheating was wrong. The moral of the story is that even the best intentions of teachers can lead students astray in the world of examinations.

Many techniques described in both parts of this collection are designed to help reduce student anxiety, and many others can be imagined, especially with the help of postcourse interviews of students. As faculty report, some of their innovations clearly result in more accurate assessment of student performance. But ultimately the only answer to better assessment of student competence on a broad scale will be more resources and more time.

GEARING UP FOR CHANGE

Mary Nakhleh, Professor of Chemistry, Purdue University

The triadic model, designed by Dr. Abigail Lipson of Harvard University's Bureau of Study Counsel (see "Commentary" by Abigail Lipson, p. 127), rightly indicates that examinations are integrally connected to the goals and curriculum of a course. The "what," "how," and "how well" elements of the model relate to content, curriculum, and assessment respectively. However, the "how" element must also include the type of instructional practices used. Different forms of assessment require new models of instruction unlike the traditional lecture mode. In fact, although the triadic model nicely allows that instructional practices are part of the "how" element, the importance of "how" can be overlooked in the effort to change the "how well."

Lecture is an efficient mode of knowledge broadcasting that goes well with multiple-choice questions that test simple recall of facts or skills in applying well-known algorithms. However, if examinations become more open-ended and probe for understanding of concepts, then total reliance on lectures and traditional problem sets becomes more problematic because *broadcasting* knowledge is not the same as *transmitting* that knowledge to the learner. In fact, a large body of research now indicates that learners do not simply absorb understanding of a subject; instead, they must construct that understanding for themselves from texts, teachers, peers, common sense, and their real-world experiences.[1] A corollary of this constructivist approach to learning is that students' understanding of science concepts has depth in some areas, gaps or weaknesses in others, and sometimes outright misunderstandings. Therefore, it is appropriate that some professors are trying alternative forms of assessment not just to judge them but to uncover how students have constructed their understanding of a topic. If students are to be assessed by exams that probe their understanding of concepts more deeply, they will need to spend some class time interacting with each other and with their professor in order to debate their understanding with each other and perhaps change their weak ideas.

Constructivism, used here in the context of Lipson's triadic model, supports changing instructional practice so that students have an opportunity to develop their understanding. These changes are part of a larger national discourse on the goals of college-level education in the sciences. These goals include professional training, societal awareness, and personal enrichment. These diverse and equally important goals require that professors consider carefully several factors in planning a course: what goals to emphasize, how those goals influence the content to be taught, and how the method of assessment influences the way in which the content is taught. In terms of the triadic model, the what element must lead us to reflect on the "how" and "how well" components (i.e., on how we can change instruction and how that change must be reflected in our exams).

This revitalization of science instruction calls for informed and systematic change in instructional practice. Instructors must make one or two key changes, evaluate the results, refine their changes, and then implement the revised changes.[2] This is a model based on the theory of action research, and it will serve us well in this worthwhile enterprise.

SYSTEMIC BARRIERS TO WIDESPREAD CHANGE

Susan Nurrenbern, Professor of Chemistry, Purdue University

Anyone can make significant changes in the way he/she tests what students know—the significant changes can be relatively small, such as modifying one

---

[1] R. J. Osborne and M . C. Witrock, "Learning Science: A Generative Process," *Science in Education*, 67, (1983): 489–508; . von Glasersfeld, "Knowing without Metaphysics: Aspects of the Radical Constructivist Position," in F. Steier, Ed., *Research and Reflexivity* (Newbury Park, CA: Sage, 1991).

[2] S. Kemmis and R. McTaggart, *The Action Research Planner*, 3rd ed. (Victoria, Australia: Deakin University Press, 1990).

or two questions per exam or changing a quiz format from individual to cooperative work. However, widespread change in testing strategies in science classes will not occur for some time. One reason is that pockets of innovation and change are susceptible to personnel changes. Another is institutional resistance to change because of rigid structures of faculty governance, and the resistance on the part of the majority of noninnovative professors that they be given "academic freedom" to determine the shape, nature, and format of their teaching. Therefore, changing testing strategies may remain only a kind of hobby, no matter how passionately pursued, for personal satisfaction and the thrill of achievement.

Faculty tend to hide behind the *status quo*, arguing that there is insufficient institutional support for major reform. But, when institutions (e.g., presidents, deans) do take on leadership roles in teaching and learning activities that involve changes in instruction, instructors object. They feel their academic freedom has been violated, and that it is "unfair" to reward certain instructors and not others for making innovations. Others are unwilling to support innovation or change, because they think there will be more expected of them if the change is made mandatory.

In recent years, the widespread use of student evaluations, introduced in response to calls for greater accountability, has inadvertently created yet another barrier to course improvements. Instructors say they are unwilling to undertake anything that might be "potentially upsetting" to students, such as a radical change in the type and format of exams.

Despite the unlikelihood that innovation can lead to widespread change in the near term, there is no reason to be pessimistic about the future. The hope for long-term change lies in the new generation of instructors being educated. If they experience alternative modes of testing in their science classes and in their training as TAs, they may be more comfortable with, and knowledgeable about, incorporating different testing strategies into their classes.

REVISING A TRADITIONAL CURRICULUM

Arthur B. Ellis, Professor of Chemistry, University of Wisconsin–Madison

As noted elsewhere, the key question that should be raised in rethinking examinations is what knowledge and skills students should take away from the course: In business-quality terms, the question becomes, "What's the value added?" In scientific parlance, it is "What's the delta?"

To my knowledge, the chemistry-teaching community has yet to create a checklist of course knowledge and skill objectives. My personal list is geared to providing a "big picture," inspiring a perspective about chemistry and the development of skills that can be used throughout a student's lifetime to ensure continuous productivity. Examples include developing comfort and facility with atomic/microstructural and macroscopic descriptions of matter, making and interpreting graphs and tables, visualizing in three dimensions, seeing analogies across chemically based systems, improving verbal and written communication skills, cooperating on problem solving with classmates, and embracing interdisciplinary problem solving.

Once the benchmarks have been established, the examination is ideally a tool that helps teacher and student assess progress in attainment of the knowledge and skill objectives. In retrospect, my traditional multiple-choice based examinations did not communicate the proper message. They were commonly viewed as either rote plug-and-chug exercises and/or as efforts to trick students, leading to an adversarial student–teacher relationship. Furthermore, the fact that the exams were graded on a curve, that considerable memorization was needed, and that there was a fixed time limit led to considerable test-taking anxiety for many students.

In attempting to address these problems, a model that was useful to me at the time was the graduate student oral examination. Typically, a small group of faculty examiners will ask questions calling for insights and application of concepts from different areas of chemistry during the exam. Moreover, if a student is stuck when answering a particular question, they provide hints and generally try to make the student as comfortable as possible, recognizing how intrinsically stressful such an exam may be. Although it is impractical for me to give individual examinations to the several hundred students in my general chemistry course, I looked for a way to make the experience less stressful, to encourage more creative thinking, and to introduce options for guided thinking into my examinations.

Many of the contributors to this book have found creative solutions to the problems with traditional examinations, solutions that are tuned to the contributors' institutional parameters. This is an important step in the right direction. Our science courses should be moving targets, in much the same way that the research enterprise is a moving target: New advances should be incorporated both into the content of our courses and into our examination measuring sticks. Computer technology, as the authors discuss in the introduction to Part 1, may soon give us many new ways to assess student progress. We should be prepared as a community to embrace these opportunities. But as the contributions to this collection so compellingly demonstrate, we must not lose sight of what we are measuring and whom we are serving.

HOW TO USE THIS BOOK

What follows is the heart of this book: the descriptions by faculty members of their in-class exam innovations generated by our solicitation. Because higher-education faculty organize themselves and their courses by discipline, their entries are organized by discipline and subdiscipline in alphabetical order by faculty members' names.

The Index is meant to help the reader locate exactly the innovations he/she is interested in—by type, by discipline, by size and nature of course (for majors, for nonmajors, and so forth.).

If this collection performs a useful function, it is our hope that others in the science education community will continue to collect and publish examples of not just what faculty members are trying to do in the way of innovative in-class examinations, but also what problems in testing they are trying to solve.

INDEX

BIOLOGY

CHAPTER 2

BIOLOGY

MODIFIED MASTERY SYSTEM

Terence R. Anthoney, Department of Zoology, Southern Illinois University at Carbondale, Carbondale, IL 62901-6501; TEL: (618) 529-1660; FAX: (618) 453-5861; E-MAIL: anthoney@siu.edu.

Courses Taught

- Zoology 480—Research Methods in Animal Behavior

Description of Examination Innovation

Unlike traditional courses in physical science, many biology courses do not have problem sets that allow students to assess their mastery in depth, chapter by chapter. To remedy this perceived lack while teaching principles of research in a largely self-instructional mastery format, Terence Anthoney devotes most of his class time to practice exams, followed by group discussions of students' answers.

Anthoney's research methods course is divided into six units: definition of a research problem, research design, observation and description (of behavior), data analysis, information retrieval (library skills), and article-critiquing. Prior to each unit, he provides students with written objectives describing what he expects them to be able to do after mastering the unit. Wherever possible, Anthoney provides criteria that would demonstrate mastery of a particular objective. For example, the objective for the data-analysis unit reads as follows:

> When presented with data collected on one or more organisms, appropriate research questions and/or test results, you will be able to describe the population to which you would feel "most comfortable" generalizing the test results; and you will be able to give your reasons for generalizing as far as you did and for generalizing no further.

The practice exams in each unit focus on those objectives that students traditionally have most trouble answering. For exams on research design, Anthoney gives students actual research problems to flesh out. Students must defend their decisions as to selection of subjects, sampling methods, and length of study. For exams on observation and description, Anthoney uses motor patterns modeled by "live" humans and filmed animals, which allows multiple repetitions upon demand. He also gives students sample descriptions, which they critique for ambiguities, internal inconsistencies, and incompleteness.

For practice exams on data analysis, Anthoney gives the students actual data sets from which they must formulate and test research hypotheses. He also provides a variety of written statistical procedures, with equations and tables, so that students need not memorize them. To test ability to critique an article, Anthoney gives the students an open-book exam in class after they have already read the article as often as desired over several days and have had a chance to receive answers to their questions about the article. They are thus confronted with how much (or how little) of the information in the article they are able to abstract—a real eye-opener and motivational tool for most students.

Grading for Anthoney is a form of certification: What does each letter grade earned tell the instructor about how much a particular student has learned? A, B, C, D, and so forth, represent decreasing levels of mastery. For each real exam, Anthoney works out the criteria for the assigning of every point and determines the cutoffs for each grade ahead of time. Anthoney recognizes the "subjectivity" in his grading system, but he stands by it because it's not capricious and it's fair. "I can always translate what an A in my course means," he says. "The standards remain consistent throughout the years." A typical final-grade distribution in the class is 40 percent As, 40 percent Bs, and 20 percent Cs. The instructor points out that this may be slightly narrower than what is found in comparable upper-level courses because he begins his practice evaluations right away, and students who don't do well drop the course quickly.

Anthoney's classes have no more than 12 students. Since group discussions are so integral to his teaching, the instructor considers 8 students to be the ideal size. Most of Anthoney's students are upper-level undergraduates and graduate students.

The instructor notes that most students have great difficulty in three areas: learning to identify and describe systematically the myriad of variables in a given research situation; learning how to formulate pairs of null and research hypotheses correctly from a given data set and then run statistical procedures that properly test the stated research hypotheses; and learning how to read a research article critically and thoroughly enough to come up with reasonable hypotheses that rival those posited by the authors.

Though some students complain about the clarity of Anthoney's learning objectives and his ability to explain difficult concepts adequately, they are generally positive about the exams. And, despite frequent comments about the difficulty of the course material, most students give the course and the instructor excellent ratings.

No other faculty in the two schools in which Anthoney is affiliated—medicine and zoology—have borrowed his testing techniques, although faculty in medicine are using self-instructional objectives and problem-based learning.

ADVANCE DISTRIBUTION OF TEST QUESTIONS

Ken Baker, Department of Biology, Heidelberg College, Tiffin, OH 44883; TEL: (419) 448-2224; FAX: (419) 448-2124; E-MAIL: kbaker@nike.heidelberg.edu.

Courses Taught

- Animal Behavior

Description of Examination Innovation

Like many of our contributors, Ken Baker tries to teach critical-thinking skills that are important to scientific investigation. Toward that end, he recently prepared a set of questions in the form of a handout for each chapter in his animal behavior course. The handout identified what he expected students to take away from the chapter: (1) identification of key terms; (2) questions about the interpretation of figures and tables; (3) questions that extrapolate from basic concepts; and (4) questions that ask student to recognize the author's (or authors') point of view.

With only nine students, Baker was able to go over questions students found difficult during his in-class preexam review. Then, he constructed his tests out of the same or slightly modified questions from the study sheets. Although the test he gave was not a particularly different instrument in terms of content and format than exams he had given in previous years, the "momentum" established by the "lead-in" handout changed students' attitudes profoundly.

"My students suddenly were responsible for understanding thought-provoking, critical-thinking questions that were central to the issues I wanted them to understand. It was no small feat to be able to answer these questions," says Baker. The new format served students' needs as well. "There is so much material in animal behavior that the presence of familiar questions reduced exam stress." Indeed, it took Baker a long time to prepare the 2- to 3-page handout for each chapter. The questions were not simply look-up-in-the-book questions. Often, they required students to recognize relationships among facts that were treated as discrete entities in the text.

Baker's new testing technique gave him a better teaching experience with this class than with any other in the recent past. His role changed from repositor of information to someone who would highlight important issues brought up in students' debates and questions. His students were responsible for raising questions and explaining their difficulties with the material, which in turn led to more substantive classroom discussion. Students knew generally what would be covered on a test, but because Baker modified the study questions, they had to truly understand the material, not memorize answers to specific questions.

Baker has since used a similar testing technique in an environmental science summer course and in a nonmajors biology class with 45 students.

OPEN-BOOK SECTION ON FINAL EXAM

Donald C. Beitz, Departments of Animal Science and Biochemistry–Biophysics, 313 Kildee Hall, Iowa State University, Ames, IA 50011; TEL: (515) 294-5626; FAX: (515) 294-6445; E-MAIL: dcbeitz@iastate.edu.

Courses Taught

- Physiological Chemistry
- Biochemistry (graduate)

Description of Examination Innovation

Donald Beitz is convinced that the open-book final examination is an excellent instrument for assessing students' comprehensive understanding of physiological chemistry. Although it is uncommon for instructors in biochemistry to use open-book exams, Beitz wants his students to learn concepts and applications of biochemistry rather than cramming lots of material into short-term memory for exams. With the open-book format, Beitz is freer to ask more in-depth questions that require students to apply biochemical information to a question they may not have encountered before. Most ask for short written answers from the student. For example, one such question asks, "Why does the mature erythrocyte of a pig produce lactate rather than pyruvate from glucose catabolism via glycolysis?"

"Because much of biochemistry involves long sequences of chemical reactions, students can get bogged down in memorizing sequences and never learn to integrate these sequences with biological and physiological functions," says Beitz.

The first half of a typical final exam covers new material from the last 10 lectures and assignments and is closed-book. Then there is a comprehensive open-book section, for which students may use any resources they wish except old exams. In preparation for the final, Beitz advises students to prepare, in their own format, a 1-page diagram of all metabolic sequences that have been discussed in class. They may bring these diagrams to the exam. In addition, students on their own initiative index their textbook and notes—some with great sophistication, Beitz notes—so it will be easier to look up details during the exam. These activities by themselves promote integration and in-depth understanding of underlying concepts, he believes. And, in this form, the notes and text will be more useful to the students in later course work.

Students in Beitz's physiological chemistry course, who often are working toward degrees in veterinary medicine and may have as many as four difficult exams in 1 week, seem more relaxed during open-book exams, according to the instructor. They also appreciate being given extra time—4 hours total—to complete the final, which is designed to take 2 hours. Students tell Beitz they like the open-book section, and news must be spreading: His students ask about it before Beitz has even broached the subject of the final with them. Still, they don't do that much better on his open-book than on his closed-book exams, in part because his open-book questions are harder.

Beitz isn't sure he'll ever give entirely open-book finals. He says there are some basic facts and principles he wants students to learn, and he's not confident they would do so if permitted to just look them up during an open-book exam.

NONTRADITIONAL PRACTICAL EXAM, DIGITAL IMAGING APPROACHES

Robert V. Blystone, Department of Biology, Trinity University, San Antonio, TX 78212; TEL: (210) 736-7243; FAX: (210) 736-7229; E-MAIL: rblyston@trinity.edu.

Courses Taught

- Microanatomy
- Developmental Biology
- Cellular and Molecular Biology
- Introductory Physiology

Description of Examination Innovation

For 25 years, Robert Blystone has been trying, as he puts it, "to better measure student performance, to encourage students to do better, and to respond to changes in how students learn" in his microscopy-based biology courses. In particular, he has improved on an old standby in biology testing, the practical exam.

In the traditional practical, the instructor sets up a series of microscopes with different objects to be identified. An ocular pointer indicates a structure, and a question card asks the student either to identify or give the function of that particular structure. The student has 30–60 seconds to record an answer, then move to the next microscope station. Such practicals would traditionally engage 30–50 microscope stations, with up to 100 questions. Students have between 30 and 50 minutes to take the test.

Blystone came to realize that this format doesn't allow students to learn from their mistakes because each practical is immediately disassembled to provide space for the next lab. Nor does the setup give the student an opportunity to make comparisons or to deal with the variability of appearances and stains that a working scientist sees. For Blystone, in terms of pedagogy and science, the standard practical was a disaster.

Fifteen years ago, he started allowing his students to "wander" the practical, staying as long as needed at each station and returning to it if they wished to compare questions and images at the various stations. In addition, after the exam, Blystone would conduct a postmortem, giving those students who wished to stay immediate feedback on their responses. Blystone noticed a 5- to 10-point improvement in student exam performance, and he observed that students were more relaxed when they took the exam this way.

Next in his exploration of alternatives, Blystone started giving students a list of every structure for which they would be responsible on the practical. At the time of the exam, each student received a randomized subset of the list and was instructed to find those structures among his/her study slides within a 60–90 minute period. Working individually in groups of three, a student would put the pointer on a structure, call Blystone over, and the instructor would agree or disagree. If he disagreed, he would move the slide to reveal the correct structure. This gave the student immediate corrective feedback.

Blystone says student performance improved dramatically using this format. "Students either knew the material or didn't," and grades fell into two categories: 85–95 or 65–75. Best of all, the poorly performing student had role models for better performance. A disadvantage was that the instructor could work with only 3 students at a time. Hence, with 30 students, it took about 10 hours to administer the exam.

With the advent of computers and digital imaging, Blystone now has students "frame grab," or digitally capture, images from the microscope slide-study set and construct a "find the structures" list. While they work (still in threes, as Blystone uses four digital imaging workstations), he can be in the room working on other materials, such as grading lecture exams. When students complete their timed computer session, he quickly reviews their labeled frame grabs. Correct responses are removed from the hard drive and incorrect responses are retained and reviewed in depth with the student at a later time. Blystone also stores images for which he can ask quantitative questions that require measurements and analysis.

"This type of questioning is not possible at all with the traditional style of testing," says Blystone.

A second digital approach to testing has worked better than Blystone imagined. He gives students a specific assignment, such as to identify examples of each of the primary tissues, and they frame grab examples, label them, and use Microsoft PowerPoint, Adobe Photoshop, and other software to sequence the images into a presentation. Students have 15 minutes to take Blystone through their presentations and "impress" him with their knowledge. Blystone comments on intermediate stages of their work during the 2 weeks they develop their presentation practicals.

Blystone organizes the presentations into an automated sequence and produces a videotape of them. The tape allows Blystone to review each student's presentation for an in-depth analysis of his/her understanding while sharing the work with all students.

"This visual review of many students' work reveals different visual patterns of learning and material organization by students," he says. For example, some students build composite images with high degrees of visual impact, whereas others just digitally glue pictures together. Still others show one image at a time. Students individualize their presentations as well. One might develop strong visual content for the nervous system, whereas another might emphasize connective tissue. A few students will include quantitative elements, such as ranges of capillary sizes or neuron sizes.

Blystone says these observations allow him to guide students more effectively and help him see what he needs to emphasize in review lectures. Blystone estimates fewer than 50 colleges are using digital imaging, due to cost, unfamiliarity, and the changes introduced by the technology in the traditional classroom environment. Although these methods are challenging, Blystone says it is gratifying and enjoyable to directly participate in the students' learning.

During a recent parents' weekend at Trinity, about a third of the class brought their parents to the lab to show off their presentations. Although students usually have only 15 minutes to do the job, Blystone noted that one student held his parents in front of the computer monitor for an hour, patiently explaining to them in detail the tissue types he'd learned.

SHORT ESSAY QUESTIONS, VARIABLE WEIGHTS FOR EXAMS, UNTIMED EXAMS

Barbara Brennessel, Department of Biology, Wheaton College, Norton, MA 02766; TEL: (508) 285-8200 Ext. 5615; FAX: (508) 285-8278; E-MAIL: Barbara_Brennessel@wheatonma.edu.

Courses Taught

- Biochemistry
- Microbiology and Immunology

- Nutrition
- Molecular Biology

Description of Examination Innovation

Over the last 12 years, Barbara Brennessel has tinkered with various aspects of her exam design, format, and grading practice to encourage open-ended inquiry and collaboration. Between 30 and 70 percent of each of her exams call for short essays describing and then elaborating on a solution to a problem. Few of these questions will have a single correct answer, and multiple interpretations are encouraged. Students might be given the results of one experiment and asked to interpret them, or they may be asked to design a series of experiments to test a hypothesis. Or, they may have to comment on a newspaper article describing a new finding in biology. For example:

> "You have traveled to the Gobi desert and have isolated a new microorganism. It is a halophilic, eurythermal, facultative anaerobe. What procedures did you use to culture and isolate this new organism? What will your name it?" A nutrition question might ask students to use their "nutritional wisdom" to comment on statements such as "Everyone should eat oat bran" or "Aspartame-sweetened beverages, rather than sugar-sweetened beverages, are a better choice for children."

"This kind of question tests students' ability to apply what they have learned rather than simply to recall it," says Brennessel. She reports that although many students react favorably to this exam format, some complain that she isn't testing them "on what they studied." Perhaps they mean what they have memorized—which is just her point. They're frustrated, according to Brennessel. "Although they say they've studied, they still can't answer the questions. I tell them it's okay to be frustrated. Real science is like this," she says. But many of her students aren't used to this kind of testing environment.

On the essay section of her exam, Brennessel often finds her students' solutions novel and exciting. Because her class is small, she can do the grading by hand and by herself. She uses simple criteria: Is the science content appropriate to answer the question? Has the question been answered thoroughly? And how well developed is the student's answer? Early in the semester, Brennessel informs her students of these guidelines. Sometimes, she says, grading is straightforward; other times, she has to read all the answers first to decide where the cutoffs should fall.

Because some of her students have trouble with higher-level thinking, she dedicates 20–30 percent of her exams to questions that test definitions and factual recall. Testing on facts rewards students for studying; it also relieves some of the frustration they feel if they read through the exam and don't know how to begin. By getting them to answer some straightforward questions, they develop more confidence for the rest of the exam.

Partly because it takes time to get used to her open-ended testing format, Brennessel weights exam grades so that students' lowest grade counts least and their highest counts most toward a final grade. This method also reduces test anxiety. Brennessel gives three exams per semester worth, in total, 65 percent of students' final grades (lab accounts for the remaining 35 percent). Originally, she had students decide in advance how they'd like their test grades weighted. But when she did that, some just studied hard for the final and let the others count very little. So Brennessel stipulated that no exam could count for less than 10 percent or more than 30 percent.

"Students doing good work get good grades," says Brennessel of this method, "but if I average all the exams equally, I find that my approach doesn't affect the final grade significantly either way." Weighted grading has the advantage of making the students feel a little less stressed about exams.

To further relieve test anxiety, Brennessel has recently been giving untimed exams. When, in a recent semester she had learning-disabled and limited-English-speaking students in her class, she made all her exams available 1 hour before class and allowed students to do the exam in an empty classroom or lab. The system works at Wheaton College because there is an honor code in place. Some students spent over 3 hours on exams designed to take 1½ –2 hours. The instructor also reduced the number of questions to 10 per exam and weighed them all equally, so students could easily budget their time and energy.

For herself, Brennessel finds it challenging to make up and grade exams like this. "Any time I'm reading a news item, I put it in the file. For example, something about India's plague epidemic will be on a future microbiology exam. A real-life situation is often the basis for a question I can pose." Her goal is to get students to realize they can use what they have learned. Sometimes, she will ask them to design an experiment, or ask them how they would approach a problem.

At first, Brennessel's students are terrified of her tests—they'd prefer a checklist of equations to memorize. For students who have difficulty understanding questions that are not of the usual, standardized type, the instructor will rephrase a question. (Learning-disabled students may orally explain or expand on their written answers.)

Some of Brennessel's colleagues are coming to agree that 1 or 1½ hours isn't enough time for the careful thinking required of exams and give students extra time. And in the smaller classes, of which there are more at Wheaton than at larger universities, faculty are experimenting with other assessment ideas. "Most of us think it's important to get students to write and not just circle answers," says Brennessel.

TAKE-HOME EXAM BASED ON RESEARCH LITERATURE, CHOICE OF TAKE-HOME OR IN-CLASS EXAM

Suzanne Connors, formerly of Department of Biology, San Jose State University, San Jose, CA 95192-0100.

Courses Taught

- Introduction to Principles of Toxicology
- Biological and Systemic Toxicity
- Genetic Toxicology

Description of Examination Innovation

In her upper-division toxicology courses while at San Jose State University (which regularly enroll about 25 students), Connors developed students' reasoning skills in the context of toxicology experiments. To encourage higher-level thinking, Connors' students in Biological and Systemic Toxicity spent time learning how to read a scientific article critically. They were assigned 30 papers during the semester, and, in the "in-class journal club" activity, the papers were discussed and critiqued.

Connors also tested her students on reasoning skills. In one long-term assignment, students were given a journal article dealing with the identification of drug metabolites in the urine and feces of rats with the names of the chemical structures analyzed in the article whitened out. From data from mass spectroscopy, nuclear magnetic resonance spectroscopy, thin-layer chromatography, library references, and so forth, students had to identify the missing structures. Although they did not receive partial credit on exercises such as this, Connors's final required that students understand the functioning of different kinds of instrumentation used in toxicology research.

Most importantly, from Connors's point of view, the assessment scheme forced students to review and recall the basic chemistry and biology upon which toxicology is based. "Often, students don't understand why the basic sciences are a prerequisite for upper-division classes," says Connors. The types of questions Connors posed pull together what they'd learned before. To understand the instrumental analyses of thin-layer chromatography or mass spectroscopy, for example, a student must draw heavily on chemistry and biology from previous courses.

Exams and projects that count as much as a final shouldn't be used as a "test" or as a punishment, Connors believes. They should function mainly as learning experiences. Connors's introduction to principles of toxicology course, with 100 students, was too large a class for the long-term assignment. So in that course, she gave her introductory students six to eight optional journal articles to read. Sometimes Connors presented original data from her own experiments to reinforce concepts also presented in the articles. "I try to let these students 'peek' into my scientific professional life so that they can see the practical applications of concepts," she says.

To deal with the time stresses of in-class examinations, Connors allowed her students to choose between taking exams in class for 3½ hours each or at home, with 2 weeks to complete. (This second option bore a 10 percent penalty.) Or, they could opt for four hourly exams, each worth 25 percent of the grade. The average of the four exam grades (adjusting final scores according to class averages) became their final grade. Only if they missed an exam did they have to take a comprehensive final.

"Every year I try something new," says Connors. These options gave students some flexibility so they could prepare for her exams around their busy schedules. The options also permitted her to create exams that were more difficult than the ones her students were accustomed to taking.

Connors is aware, as she puts it, that college "is the last place students can be safely challenged without severe consequences." She wants to take full advantage of that, even if she gets "punished" on student evaluations, which she says often happened. More significant were students' long-term evaluations of her teaching and testing strategies: "Students who go into industry and Ph.D. programs come back and tell me I did the right thing," she says.

MODIFIED MULTIPLE-CHOICE QUESTIONS TESTING HIGHER-ORDER THINKING

Glen Erickson, F 205, Delta College, University Center, MI 48710; TEL: (517) 686-9260; FAX: (517) 686-8736; E-MAIL: geericks@alpha.delta.edu.

Courses Taught

- Human Anatomy and Physiology

Description of Examination Innovation

Glen Erickson's multiple-choice test questions are designed to help students learn to apply anatomy and physiology to novel situations they haven't discussed in class or read about in the textbook. Each "situation" is followed by three or four multiple-choice questions that assess different aspects of the "problem."

For example, students are told that a patient arrives at a hospital emergency room with gunshot wounds. X rays reveal that a bullet is lodged in the patient's right lung. Students are then given four choices to describe the probable future of this patient in terms of his respiratory system (from the patient exhibiting dyspnea until death occurs, to dyspnea and respiratory compensation) and are asked to explain which outcome is more likely, and why.

For 10 years Erickson has been using questions like these instead of the more common identification questions used in anatomy and physiology courses (i.e., "Define dyspnea"). Most textbook questions test what Erickson, echoing William Perry,[1] calls "first-level" thinking: regurgitation of facts. "Second-level" thinking, in his view, requires that students put facts together to answer a question. Some degree of synthesis is always required.

"Third-level" questions—the ones Erickson likes best—require students to know the anatomy and physiology of entire organs and their place in the body's system as a whole. Absent this level of mastery, Erickson predicts, a student won't even understand the question. Furthermore, the instructor's questions require that students deal with some unique situation that has not previously been covered in class. Although no partial credit is given for third-level questions, students can answer one question in a series correctly while missing the others, as when they know the anatomy but not the physiology of a certain situation.

Student reaction to these questions has been generally positive, although some students say the questions are "really hard" or "unfair." In Erickson's view, reading difficulties contribute to a reduced success rate with these questions.

Delta College holds an occasional faculty seminar on testing. Some years ago, Erickson reported to his colleagues on his testing mode. Years earlier, the had gathered questions like his for the general biology final but, in time, began to worry that they were relying too much on reading ability and abandoned the exam.

Even so, Erickson has stuck with his testing strategy. His classroom research indicates that "situation" questions, when carefully constructed, evaluate students' critical-thinking skills just as well as essay questions. Not that any instructor can actually teach critical thinking, says Erickson. But he believes continual use of and review of the skills involved in teasing out factual material from complex situations will encourage critical thinking among students.

INTERPRETIVE EXERCISES

Jinnie Garrett, Department of Biology, Hamilton College, Clinton, NY; TEL: (315) 859-4716; FAX: (315) 859-4807; E-MAIL: jgarrett@itsmail1.hamilton.edu.

Courses Taught
- Introductory Biology
- Genetics

---

[1] William G. Perry, Jr., *Forms of Intellectual and Ethical Development in the College Years: A Scheme* (Troy, MO: Holt, Rinehart & Winston, 1970).

- General and Molecular Genetics
- Gender and Science

Description of Examination Innovation

In her upper-level general and molecular genetics course, Jinnie Garrett has replaced exams with what she calls "interpretive exercises," a graduate-level assessment format involving primary literature. Although her technique requires significant class time so that students (mainly juniors and seniors) understand how to critique the articles they read, Garrett says it teaches students how to understand biology from primary sources.

During the class before an exercise, Garrett gives her students the introduction, methods, and results sections of a recent paper on a topic related to the subject under discussion. Articles from *Science,* the *Journal of Biological Chemistry*, and other journals are used. Students have 2 days to study the paper and determine its conclusions. They may use any source they want, except they may not query any member of the biology or chemistry faculty, nor any similarly qualified individual.

"I also encourage them to work together and help each other by discussing the paper without me being present. The first year I did this, they just ran a class and worked through the paper together," says Garrett.

During the in-class exam, students must answer a set of questions on the paper. They may use any resources and notes they have brought to the classroom. But in only 50 minutes, students usually have to rely on the understanding they've already reached, especially because the questions probe in-depth comprehension of the science in the articles. For example, one 50-point part of a three-part question (worth 100 points in total) asks students to "summarize the evidence that convinces *you* that the amino-terminal 16 amino acids of thiolase are the signal for localization of the protein to the peroxisome in *S. cerevisiae.*" Questions like these also require students to explain in their own words the experimental methodology of the experiments, as well as the reasons for their success or failure. Sometimes students are even asked to step into the researchers' minds and explain why they might have certain premises. Part of their challenge is to evaluate the articles, explaining how an experiment could have been made more convincing.

To reduce stress and anxiety, Garrett counts only the best three out of six responses to interpretive questions. "I really like this format," she says, "because it encourages students to work in teams and expand the frontiers of their understanding both collectively and individually. It also gives me a clear-cut individual effort to grade at the end. I don't have to worry about who did most of the work because I'm grading one student's understanding of the paper."

Orienting students to the new types of exams by making them aware of and comfortable with the instructor's expectations is important, says Garrett. To this end, she spends only one-third of class time on traditional lecture; the rest involves in-class discussion of journal articles or of ethical and social issues in science. Thus, in a single semester, 11 class meetings will be spent analyzing primary literature, including the background of the experiment described, how (and whether) the experiment adequately answered the research questions, and whether the right controls were present. Her students' confidence grows as the semester progresses.

Students appreciate not having to memorize facts out of context and, according to Garrett, seem to gain a real sense of accomplishment from the process. Garrett's teaching evaluations have been uniformly positive for the 4 years she has been using this technique. Students have said they find the course "very demanding," with its high standards, but there have been no negative comments about the testing method.

MODIFIED TRUE–FALSE QUESTIONS, COMPUTERIZED EXAM DELIVERY AND SCORING, PENALIZING GUESSING

John Gwinn, Department of Biology, University of Akron, Akron, OH 44325-3908; TEL: (330) 972-7160; FAX: (330) 972-8445; E-MAIL: gwinn@uakron.edu.

Courses Taught

- Pharmacology
- Anatomy and Physiology
- Comparative Vertebrate Morphology
- Introductory Human Physiology

Description of Examination Innovation

For the past 10 years, John Gwinn has created and graded exams that reflect his learning philosophy: true understanding means knowing something well enough to stand by your answer on an exam.

Because his anatomy and physiology courses have so many students, Gwinn has had to modify traditional exam formats rather than use essay exams. He uses standard true–false questions, but in an adapted format to discourage students from thinking "in absolutes and generalities" and to offset the 50 percent guess rate. He groups together five statements, four of which are true and one of which is false, or four of which are false and one of which is true. The student chooses the one statement that is different. In this form, the true–false question requires some of the synthesis and deep analysis a well-constructed essay question demands but is much less time-consuming to grade, says Gwinn.

Although students feel they're "getting a break," says Gwinn, statistics reveal that these questions, when well constructed, are extremely discriminating. In fact, Gwinn maintains that well-designed true–false questions are challenging when they test cause-and-effect relationships.

Gwinn also uses multiple-choice and straightforward true–false questions, but to discourage guessing, he scores exams by the number right minus the number wrong, with unanswered items not included in the score at all. Gwinn adapted this grading practice from a former professor of his not only to penalize guessing, but also to remind the class that deep understanding of the material is the most important goal of the course.

To further underscore the importance of conceptual understanding, Gwinn gives students the opportunity to qualify their answers with a written justification. Therefore, if they've interpreted an ambiguous question in a unique way, Gwinn can give them at least partial, if not full, credit.

On a test consisting of 25 items, there are an average of two to three justifications per exam, per student. About half the time, these students get some credit for answers they marked incorrectly (because Gwinn sympathizes to some agree with their interpretation), and about half the time, their justifications don't bear on the answer, and the score remains the same. Occasionally, a correct answer accompanied by an inaccurate explanation is marked wrong.

Gwinn's "odyssey" in assessment began about 15 years ago with an on-line computer testing program he designed for large anatomy and physiology classes. It was one of the first computerized testing programs to combine testing and learning functions with special feedback for incorrect and correct answers.

Although he no longer uses this computer testing model (in part because the class was canceled by the department), other departments picked up on the model. What isn't being retained, at least to Gwinn's knowledge, is the extensive feedback loop, in part because it is so time-consuming to design the feedback for all the distractors. Today, in computerized testing, students are given immediate feedback as to whether they've answered a question correctly. But Gwinn's model was designed to provide more—a sentence reinforcing the connection (showing a new relationship or using different terminology) between the question and the correct answer if the correct answer is given and the reason why the choice is incorrect if a student answers incorrectly. With Gwinn's software, the student could attempt the same question again and receive half credit if the second answer was correct.

Gwinn says his system greatly reduced student anxiety about computerized testing due to both the half-credit option and the help students received in the process of taking the test. Without this feedback loop, he believes computerized testing, at least on his campus, is a less effective teaching tool.

OPEN-BOOK ESSAY EXAMS ON ORIGINAL SCIENTIFIC SOURCES, GROUP ORAL PRESENTATIONS

Judith E. Heady, Department of Natural Sciences, University of Michigan–Dearborn, 4901 Evergreen Road, Dearborn, MI 48128-1491; TEL: (313) 593-5477; FAX: (313) 593-4937; E-MAIL: jheady@umich.edu.

Courses Taught

- Introductory Biology
- Histology
- Embryology
- Comparative Anatomy of Vertebrates
- Gender and Science

Description of Examination Innovation

Judith Heady uses many different assessment strategies with students, including multiple-choice questions that must be defended, journals, and class discussion. However, two techniques stand out.

Heady teaches students how to read research articles and other scientific evidence with a critical eye by using in-class essay exams. For her women's studies and biology exams, students are encouraged to bring to class whatever books, journals, research papers, news articles, and class notes they wish and are told to choose one of between three and five questions to answer in the hour and a half allotted for the class period. To prepare for the exam, students have been given review questions delineating the scope of the test and providing a reasonable review for at least one of the questions on the exam.

One of the questions from the first examination in embryology illustrates how she incorporates research and critical thinking into a question on autonomous development and induction:

> What is the difference between induction of cells and autonomous development of cells in embryos? Describe two or three experiments with evidence that clearly shows the differences. It will be best to use at least one example of each cause of cell change and then a third, if desired. You may supplement your answer (extra credit) with a description of a hypothetical experiment to differentiate the two causes of change.

In embryology, instead of using a textbook, students read original research articles collected by the instructor. Students also have access to other articles on the same topics. To help students learn how to read research papers, students in Heady's class give presentations and conduct small-group and full-class discussions on the material in the articles.

She grades these essays on clarity, whether the response answers the given question, use of appropriate examples of research to support conclusions, effective discussion of the experimental work, and linkage to other ideas discussed in class. Heady will award extra credit to students who propose new studies to further the research.

"My emphasis is on having students understand how and why these studies were done, and why the researchers excluded other explanations—if they did," she explains. As a result of this practice, students become quite critical of experimental methods and conclusions.

Heady will award students extra points for "good ideas," even if the writing isn't up to par. Likewise, extra points are given for particularly clear evaluations and presentations, even if the ideas aren't strikingly original.

Students in Heady's biology classes have been giving group presentations for the past several years on subjects linked to animal behavior, genes and disease, and other topics. Students must prepare an outline, connect the presentation to topics covered in class, and make clear, interesting, well-researched, and up-to-date presentations. For example, for the last assignment on animal behavior in comparative anatomy of vertebrates, a group of three students combined their interest in human development, medicine, and vertebrate anatomy to make a presentation on the evolution of human walking. Their presentation included drawings of the changes in the hip areas, forearms, and shoulders of humans and other primates, and demonstrations of human walking and comparisons with observations made at the zoo of other primates in motion.

Students evaluate themselves and each other, which Heady uses in her written assessment of the presentation. Although there may be some variation, a group grade is given. Heady's students say they like this part of the course better than any other because they can be creative with videos, posters, handouts, computers, and even occasional costumes.

For one embryology student, understanding research papers became easier with practice, even when the papers got more difficult. The student came to realize how much she'd increased her skills. And when Heady asks students a few weeks into the course if they want to continue with open-book essay exams using research, they say they do not want to go back to memorizing details. Heady has also received favorable comments about group presentations of original research.

Some natural science faculty at University of Michigan–Dearborn use group presentations in their classes, but Heady knows of no other who gives in-class essay exams. Other instructors are now using more variable types of assessment such as portfolios and creating a "book" on the World Wide Web.

COLLABORATIVE TAKE-HOME QUESTIONS ON FINAL EXAM

Anne Heise, Department of Biology, Washtenaw Community College, 4800 E. Huron River Dr., P.O.B. D-1, Ann Arbor, MI 48106; TEL: (313) 973-3363; FAX: (313) 677-5414; E-MAIL: aheise@orchard.washtenaw.cc.mi.us.

Courses Taught

- Botany
- Microbiology

Description of Examination Innovation

A portion of Anne Heise's microbiology final examination includes questions to be answered collaboratively outside of class. One of her most innovative questions asks students to design a 10-week course in microbiology for a given audience, such as nurse's aides or home-health aides. Students select the topics to be covered and write a paragraph outlining each lecture—not the facts so much as a statement of the material to be covered and an explanation of why this material is important. Lecture topics students commonly choose include antibiotic resistance, hospital-acquired infections, immunity, and classification of microorganisms.

At best, the assignment causes students to identify the core concepts of microbiology and then decide how to organize the material in a 10-lecture sequence. Least satisfactory is when the question drives students back to the same topics Heise outlined in her syllabus.

Most of the students in her course will have careers in health care. Hence, it is instructive for them to figure out for themselves what everybody in health care needs to know and become, as Heise puts it, "champions" of this material. Students generally like the assignment, not least because most of them get full credit. The instructor values it because the process of preparing the lecture topics—of taking on the role of the teacher—naturally leads her students to be reflective and analytical. Heise feels an in-class exam wouldn't give students enough time to answer these kinds of questions well. When they collaborate outside of class, students spend more time discussing the content of microbiology than they would otherwise.

GROUP ORALS

Alan R. Holyoak, Department of Biology, Manchester College, North Manchester, IN 46962; TEL: (219) 982-5307; FAX: (219) 982-5043; E-MAIL: arholyoak@manchester.edu.

Courses Taught

- Invertebrate Zoology
- Marine Biology

- Principles of Biology I
- Freshwater Biology

Description of Examination Innovation

Alan Holyoak noticed that his invertebrate zoology students worked harder, covered more material, and studied longer when preparing for his oral exams than for his written ones. Two years ago he tried *group* orals for the first time. He noticed an improvement in performance between the first and second group oral exams and has continued the practice in his invertebrate zoology courses, as well as in a seminar course entitled "Topics in Invetebrate Zoology."

Students are given a list of exam questions to study at least 1 week in advance of Holyoak's exam. On the day of the exam, students draw the question they must answer from a hat. The questions are numbered, and the students arrange themselves in a circle in order of their question number. Using a random-number generator, Holyoak selects the first student to answer his/her question.

The first student answers the question without interruption. Once finished, this student cannot add anything to the completed answer. Next, the student to the right of the first has the opportunity to contribute, and so on, until all students respond. One round of questioning is completed after each class member has started off his/her primary question and the others have added their comments.

Holyoak tries to control the "inherent level of subjectivity" involved in grading these presentations by quantifying responses fairly precisely. He rates the first person's response to a question in one of the following categories: "good" (9 points), "okay" (6 points), "weak" (3 points), or "wrong" (0 points). Other responses (after the first person's in each group) are rated as either "important" (2 points), "applicable but not central to the question" (1 point), "not applicable" (0 points), or "wrong" (−1 point).

With this scoring method, Holyoak generates a total point score for each student that he uses as a basis for evaluating performance. "It's a relative scale, of course, but it is very helpful in explaining the results of the exam to the students," reports Holyoak.

Depending on the size of the class and on the depth and completeness of answers, two or three rounds of questions may be completed during a group oral exam. Of course, this kind of examination works only when the class is small. In a class of 12 students, a single round of questions takes nearly an hour to complete. Therefore, Holyoak schedules group oral exams during the class's 2-hour lab period. He doesn't think group oral exams would work well with groups larger than 15 students.

Holyoak has received mixed responses to this method of testing. Almost invariably, students feel worse about their performance after the oral exam than they normally feel after taking written exams. They also feel pressured. On the other hand, once their grades are in, students frequently are pleasantly surprised by their scores. In one postexam feedback session, students said they had studied harder and learned more preparing for the oral than for the written exams. And some students were so favorably impressed by the learning-related results they gleened from preparing for and taking group oral exams that they asked Holyoak (while enrolled in other courses he teaches) if he would use the group oral exam method again.

Holyoak's colleagues think the method is interesting and reasonable, though none have adopted it, so far as Holyoak knows.

COMBINATION IN-CLASS/TAKE-HOME EXAM

James B. Johnson, Division of Entomology, University of Idaho, Moscow, ID 83844; TEL: (208) 885-7543; FAX: (208) 885-7760; E-MAIL: djohnson@uidaho.edu.

Courses Taught

- Insect Identification
- Systematic Entomology
- Biological Control

Description of Examination Innovation

James Johnson gives combined in-class/take-home midterm exams in systematics and biological control. Of the approximately five essay questions on each 90–120 minute exam, students complete two or three during the exam period and take the others home without penalty. Students are allowed to confer on the exam.

"I hope the take-home feature reinforces their learning in areas where they weren't comfortable enough to answer the question in class," says Johnson. The idea of the combination in-class/take-home came to Johnson during a discussion with students about their preferences for exam type. When no consensus could be reached, the advantages of a "compromise" occurred to Johnson.

Some of Johnson's students appreciate this combination as a way of displaying their knowledge; others dislike the added work. The instructor suspects that the added grading time discourages other faculty from using this method.

Johnson also gives oral final exams in systematics, a graduate course. He says oral exams work well in dealing with complex, theoretical topics because students can be guided back on track if they get on a tangent. Using a standard list of questions as starting points reduces variation between exams.

FINAL EXAMS BASED ON COMMON DISEASES

Carolyn K. Jones, Department of Life Science, McCormick Science Center, Vincennes University, Vincennes, IN 47591; TEL: (812) 888-4136; FAX: (812) 888-4540; E-MAIL: cjones@vunet.vinu.edu.

Courses Taught

- Human Biology
- Plant and Animal Biology
- Anatomy and Physiology
- Microbiology
- Molecular Biology

Description of Examination Innovation

Recognizing that nonmajors in biology may not want to concern themselves with abstractions, but rather, will benefit from an ability to identify the symptoms of a disease, Carolyn Jones bases her nonmajors' human biology final exam on the biology of common diseases. She has used this format for 5 years and finds that her students are more engaged in the material—and learn more.

Over the course of the semester, students in Jones's Human Biology course prepare 30 index cards describing diseases that relate to basic biological information they are discussing in class. A discussion of atoms and ions, for example, will require a card on calcium imbalance. A card on muscular dystrophy is assigned for the musculoskeletal unit, and a card on Alzheimer's accompanies the unit on the nervous system. Students' cards must include the formal biological definition of the disease, its symptoms, its current treatment, its transmittable characteristics, and the social implications for each disease. Index cards are graded for accuracy and returned. Prior to the final, Jones conducts a review during which she covers all information that is included on students' cards.

Questions on the final exam incorporate lecture material relating to the diseases, as well as information contained on the card. For example, a final exam question relating to calcium imbalance would read, "Calcium imbalance can be related to (a) muscle cramps; (b) blood pressure changes; (c) loss of appetite; (d) loss of bone mass; (e) all of the above." A question on muscular dystrophy would read, "What protein is lacking in the muscle of individuals with muscular dystrophy? What is the correlation between this protein and calcium?" A typical test question from the unit on the nervous system would ask the student to correlate the neurotransmitter missing in the brains of individuals with Alzheimer's and symptoms associated with the disease.

"I have found that when relating biology to the world of disease, the students and I have a common ground on which we can communicate. They may not previously have had an interest in the biological nature of the disease, but they are interested in the disease itself," says Jones.

At the end of every semester, Jones asks her students how useful the disease cards have been. With the exception of only two responses over a 5-year period, all students reported that the disease cards favorably supplemented the coursework, and that the information they learned was interesting and beneficial. She has also had several students approach her later in their undergraduate education and comment on how often they have had the opportunity to expound on the information they learned as a result of the disease cards.

Other faculty have reacted positively to this method. It has been used by a colleague of Jones's at Indiana State University with favorable response.

COMBINATION IN-CLASS/TAKE-HOME FINAL WITH GROUP WORK

A. Krishna Kumaran, Department of Biology, Marquette University, Milwaukee, WI 53233; TEL: (414) 288-1478; FAX: (414) 288-7357; E-MAIL: kumarana@vms.csd.mu.edu.

Courses Taught

- Cell Biology

Description of Examination Innovation

In his junior-level cell biology class, A. Krishna Kumaran gives a comprehensive final exam that combines in-class and take-home testing methods. Thirteen short essay questions are handed out to students 3 weeks before the final is scheduled. Kumaran tells students he will select some of these questions for the exam, instructing them to prepare all the questions either individually or in groups, using whatever resources they wish.

Kumaran's questions require students to integrate science content from several units. For example, one question asks how the Na/K ion channel protein is synthesized and integrated into a membrane. To answer the question correctly, the student needs to have learned and synthesized information about regulation of transcription, translation, posttranslational modifications, and how the membrane is organized. Another question asks students to discuss the evidence based on conservation of structure and/or function of cell components that all organisms must have evolved from a common ancestor. An accurate answer requires synthesis of knowledge about the conservation of adenosine triphosphate synthetase (ATP synthesis), DNA and RNA polymerase (DNA replication and transcription), codon usage and translation machinery (genetic code and protein synthesis), and the singularity of the fundamental processes of life.

Many of these questions require too much synthesis, according to Kumaran, for a clearly organized, 200–250 word answer during a 2-hour in-class exam. Pressed for time, students' responses would be schematic, summarizing or outlining the relevant science rather than truly synthesizing the material into a well-developed response. Therefore, the instructor asks these comprehensive questions in the take-home portion of the exam.

Approximately a third of all of Kumaran's exams, including the final, test students' ability to analyze the significance of experimental protocols and/or observations. In these exams, students are asked to explain the basic question the experiment is designed to answer or assess the significance of the observation in reference to the basic question. "Students must learn the process of scientific research, not just the findings," explains Kumaran.

Kumaran has used this method for 25 years and finds that it compels students to really learn the material. "Many students tell me after the test that they understood the material better than if they took a traditional in-class exam," he reports. Faculty colleagues are reluctant to use this method, fearing it will not be a "true test" of the students' learning, says Kumaran.

So long as the stress of a "cold" exam—one with questions for which the students may not be prepared—is considered the only "fair test" of students' learning, Kumaran thinks innovations like this one will remain marginal in college-level science.

POSSIBLE 110 PERCENT SCORES, OPEN-BOOK OPTION HALFWAY THROUGH EXAM, MAKE-UP EXAMS

Judi A. Lindsley, Biology Department, Lourdes College, 6832 Convent Boulevard, Sylvania, OH 43560; TEL: (419) 885-3211 Ext. 259; FAX: (419) 882-3786.

Courses Taught

- Anatomy and Physiology
- Microbiology
- Vertebrate Zoology

Description of Examination Innovation

Judi Lindsley uses grading methods and exam schedules that may be particulary well-suited to premed and other competitive-track classes in which students are concerned about grade-point averages.

In her typically large courses, Lindsley includes multiple-choice questions on at least a portion of her exams because she is trying to prepare her premed and nursing students for their board exams. However, she maintains that it is "virtually impossible to write multiple-choice questions that are entirely unambiguous." (This despite the fact that she chooses to submit her tests to a colleague's review for ambiguous questions and distracters.) To compensate, she gives students the opportunity to earn a score of 110 percent on her exams. These aren't extra credit or bonus questions, just enough points to add up to a more than perfect score, much like an A in an honors course counts more than an A in a regular class.

In the last 5 years that she's employed the 110-percent method, Lindsley says students feel they're getting a break, and she is more comfortable demanding, if a student complains about a test item, that the student justify his/her answer "on the spot" by discussing the disputed answer directly with her. Lindsley also uses other types of test question formats, such as short answer, because she claims some students are more verbal and less successful with multiple-choice questions.

Recently, Lindsley tried another test innovation in her smaller vertebrate zoology course with mixed results. Because of inclement weather in northwestern Ohio, the school closed for a time. Unable to cover some important material, Lindsley offered an open-note/open-book option on her exam—and with a twist: She offered it halfway through the exam. For the first half of the test period, the exam was closed book. Then, halfway through, students were given the option to continue with or without the aid of their notes and books. If they opened their notes and books, they had to finish the exam using another ink color. To compensate students who chose the closed-book option, she awarded full credit for correct answers in the first color and half credit for correct answers in the second color.

Lindsley's students seemed to prefer half credit to no credit. But since health professionals are expected to know the material well enough to be able to apply it in a stressful medical environment such as the emergency room, the instructor is more inclined to use the technique in her introductory courses in the future.

The "rules" of Lindsley's class are spelled out at the start of the course and serve to introduce students to the kind of real-world trade-offs and time management they will have to face on the job. To encourage students not to miss exams, she insists that makeups are scheduled during finals week, when students in rigorous science programs almost always have to cope with multiple exams. Whereas Lindsley used to have excessive absences during exams, she now reports no more than two absences on test dates each semester.

This innovation was suggested during a biology department brainstorming session to figure out what to do about makeup exams. Shortly thereafter, Lindsley discussed her method at a collegewide faculty meeting in which the same topic was discussed. Colleagues seemed interested in her method, although she and a chemist are the only two she knows of who are using the technique. Overall, the practice has succeeded by decreasing student absences on exam dates, although it has raised other issues among some faculty members.

Lindsley's method essentially gives students the option of postponing an exam until finals week if they are not prepared to take it on the designated test date—and this did not escape the attention of her students. Last year, she was confronted by a group of students who complained about the makeup-exam policy. "We did poorly, but at least we showed up," they said. In competition with one another, they noted the leverage built into the system that actually discouraged students from coming in when they weren't fully prepared.

Recently, Lindsley allowed students to take a makeup exam if they did poorly the first time around—so long as they also agreed to do it during finals week. "If mastery of the material is what exams are about, then this is fair," she says.

Although uneasy at first about her decision, Lindsley now says her assessment ideas have matured through a deeper understanding of her students and the culture in which they are immersed.

Cheating, particularly verbatim copying of lab reports, has also been the subject of a number of faculty meetings. Lindsley counters it in two ways. First, like many faculty with large classes, she scrambles computer-generated exam questions so the answer sheets are difficult to compare during an exam. Second, if two or more students are caught cheating, they "share" the grade. For example, if two students turn in exactly the same lab report, graded 80 percent, student A receives 40 percent, and student B receives 40 percent as well.

STUDENT-GENERATED PROJECTS IN TEACHER PREPARATION

Bonnie Lustigman and Ann Marie DiLorenzo, Biology Department, Montclair State, Upper Montclair, NJ 07043; TEL: (201) 655-5107 (Lustigman), (201) 655-4396 (Di Lorenzo); FAX: (201) 655-7047;
E-MAIL: lustigman@saturn.montclair.edu.
Jackie Willis, Biology Department, Upsala College, East Orange, NJ 07109; TEL: (201) 266-7208; FAX: (201) 655-4390.

Courses Taught

- Great Ideas of Science
- Senior Seminar in Biology
- Teacher Preparation courses

Description of Examination Innovation

Discouraged by students' lack of critical and analytical skills, biology instructors Bonnie Lustigman and Ann Marie DiLorenzo decided 7 years ago to try to raise the level of reflective thinking in their courses. In that time, they have developed two innovative courses—one for K–8 teachers, the other a capstone course for senior biology majors—that do not employ traditional testing methods.

In both courses, the instructors chose to rely on a series of student projects for grading purposes rather than in-class exams.

"Many of our preteaching students have poor science backgrounds, and we wanted to increase their comfort level with science. A traditional test would not help them," says DiLorenzo. Students in the "Great Ideas of Science" course (approximately 25) work in groups of three or four (all teaching the same grade level) on "module" projects. They create hands-on activities for their students based on the "great ideas of science" unit (covering ideas such as atoms and molecules and genetics). Teachers create games and puzzles and design supporting activities that integrate many skills into the activities, such as reading, writing, and use of technology. For example, in one activity created by the students, children read Dr. Seuss's *Bartholomew and the Oogleck* and then prepare "homemade oogleck" out of flour, water, and food coloring, studying, in an enjoyable way, weight, measurement, volume, and states of matter. Abstract chemical terminology takes on concrete meaning through the activity.

Lustigman and DiLorenzo wanted to assess individuals as well as the group as a whole without having to "police" their students, which would work against their goal of increasing teachers' comfort level with science. They came upon the idea of requiring that each group of students designate roles for every individual in the group. To assist the groups in this delegation of tasks, the instructors initiate frequent, ongoing discussions about the range of skills to be included in the projects and how each individual can contribute meaningfully to the activity. Most projects end up consisting of discrete subunits, each prepared by an individual student in the group. Thus students receive an individual grade for their contribution and a group grade for the project as a whole.

"As instructors, we are responsible for the facilitation of the group process," says DiLorenzo. The "module" projects have been extremely successful and have been implemented in K–8 classrooms. Examples include using gelatin plates to demonstrate the universality of microbes in the environment, a series of faces that change from smooth to widow's peak hairlines to show genetic transmission inheritance of traits, and toothpick/gumdrop/jelly bean model-construction of atoms.

DiLorenzo and Lustigman also wanted teachers to leave the class feeling science wasn't separate from the rest of their lives. Here again, instead of traditional exams that would "test" such a transformation, the instructors apply the idea of authentic assessment: Students analyze science articles in the local newspaper and devise ways to use children's stories as a basis for science lessons.

In addition, students write short pre- and postclass summaries of the knowledge they bring to each class session and afterward, how they could apply it to their classroom activities. Essentially a variation of the "two minute paper" described in Pat Cross and Thomas Angelo's *Classroom Assessment Techniques*,[2] formative classroom assessment activities such as these may increase the likelihood that teachers will implement science into their curricula, a worthy goal for undergraduates as well.

Once the K–8 class was on its way, the instructors turned their attention to secondary-science teaching. Dan Burke, now a coordinator of the Urban Systemic Science Initiative reform at the National Science Foundation, put them in contact with Jackie Willis, of Upsala College, to form a consortium of three local institutions (Seton Hall, Montclair State University, and Upsala College) now funded to implement "Great Ideas of Science" workshops to secondary-science teachers.

Lustigman and DiLorenzo's evolving ideas about the connection between curriculum and assessment led them to decide their senior students needed a chance to reflect on their science education, both with their peers and with facilitating instructors. The instructors designed a new senior seminar for biology majors, covering environmental and biomedical ethics. The "capstone" seminar requires students to make connections between different areas of science. Because the class will be looking at different scientific issues, such as genetic counseling and its potential conflict with the right to privacy, traditional exams did not seem appropriate for the kind of reflection and evaluation for which the instructors were aiming.

"Since the line of inquiry in this course is not linear, but more of an evolving questioning process, we will know if the students have mastered the science content if they can ask thoughtful questions and give good group presentations," says Lustigman. The presentations cover issues such as population control, organ donation, and breast cancer, and are followed by individual student papers arguing both sides of the particular issue. In this way, the instructors believe they can teach science and the processes of science, emphasizing that the scientist works with ambiguities.

According to the instructors, students are enthusiastic about the projects and prefer being graded this way over paper-and-pencil exams. On the occasions when a student complains that he/she has received an unfair grade, the instructors ask the students to redo parts of the project. This allows the much-needed self-correction that leads to good critical thinking.

Colleagues have shown an interest in their approach, though few are willing to try it.

---

[2] Thomas A. Angelo and K. Patricia Cross, *Classroom Assessment Techniques* (San Francisco: Jossey-Bass, 1993).

MODIFIED MULTIPLE-CHOICE "IMPLICATION" QUESTIONS

Ann M. Muench, Department of Biology, Mary Miller Center, Danville Area Community College, Danville, IL 61832; TEL: (217) 443-1811; FAX: (217) 443-8595; E-MAIL: amuench@dacc.cc.il.us.

Courses Taught

- Human Anatomy and Physiology

Description of Examination Innovation

Ann Muench uses complex multiple-choice questions on her anatomy and physiology exams that are difficult but instructive for students. With 70 students in a class, the questions can be machine-graded like any other multiple-choice item, but they test specific understanding: Can the student evaluate statements about course material and deduce when one statement is implied by the other?

Two statements are given, such as the following:

A. In the normal human female, there is a space between the ovary and the tip of the oviduct.
B. Occasionally, an ovum will fail to enter the uterine tube and will instead remain within the pelvic cavity.

Students are given five options: (1) Mark A if only statement A is false; (2) Mark B if only statement B is false; (3) Mark C if both statement A and statement B are false; (4) Mark D if both statements are true but not related by implication; and (5) Mark E if both statements are true and statement B is implied by statement A.

In the instructions for each exam, Muench spells out synonyms for "implied by":

"Included in"
"Virtually involved, even though it is not expressly stated"
"Expressed indirectly"
"A logical conclusion of"
"Hinted at"
"A necessary consequence of"

This format forces Muench's students, most of whom are preparing for medical fields such as nursing, to apply the anatomy and physiology material taught in the course. They must also think in terms of conclusions that can be drawn based on anatomical and physiological information. In a course infamous among students for the amount of memorization required, implication questions demand *more*, helping these students see that relationships among information sources are at least as important, if not more important, than mastering the facts.

Muench uses "implication questions" sparingly—a few on each of three exams per semester, both because students find them difficult and because they're hard for the instructor to write.

Since students are particularly intimidated by the implication questions at the start of the term, Muench says it's important to discuss the format with students and teach them how to approach the questions. She uses a pretest at the beginning of the semester, plus three actual previous test questions, before the first exam. With repetitive discussion, before and after each exam, about "implied by" relationships, students come to view this type of question as fair and are no longer unduly intimidated. It is Muench's hope that students will look for implications as they study.

Muench has shared this type of question with colleagues at other schools and has distributed examples of implication questions to those who are interested. She believes the implication question can be adapted for use in other science courses.

STUDENT-GENERATED EXAM QUESTIONS

Evelyn Neunteufel, formerly at County College, The Hartz Mountain Corporation, 192 Bloomfield Avenue, Bloomfield, NJ 07003; TEL: (201) 271-4800 Ext. 2553; FAX (201) 680-9878.

Courses Taught

- Anatomy and Physiology

Description of Examination Innovation

Evelyn Neunteufel has her anatomy and physiology students create exam questions, a worthy review exercise that turns out to be helpful to all, including the instructor.

Each student is told to create a multiple-choice question that requires thinking, not just recall, on the material to be covered in a particular exam. Between five and ten of the best student questions—constituting about 10 percent of the questions—are used as part of the exam. The students don't know which ones will be chosen by the instructor, but they all try hard to write one that will catch the instructor's eye and be used on the test.

Before the exam, each student writes his/her question on the front of an index card, offering five possible answers. On the back of the card, the student writes the correct answer and his/her name. This allows Neunteufel to evaluate the questions for ambiguity without being influenced by the answer chosen by the student. It also enables her to assess the value of the question for the exam without bias as to who wrote the question.

What are the advantages of this approach?

Students are forced to think deeply about the course, which helps them learn. And they're automatically motivated to do this assignment: "An incentive for students to invent good questions is that their question might be on the exam and they will already have prepared an answer," says Neunteufel. The practice may also encourage students to compare questions, thereby increasing discussion of anatomy and physiology after class.

Students report that they feel good about contributing to the exam content. This technique also helps Neunteufel formulate some good questions, ones she says she wouldn't have thought of on her own.

"Students who are professionals or working in a course-related field, such as LPNs or medics in a nursing course, invent some good job-related questions," says Neunteufel. Last year, some of her nursing students, who were working in the field, wrote questions that helped other students relate class material to their future profession.

"This method is especially valuable, because it allows students with different learning styles to teach the instructor how they approach problems," adds Neunteufel.

A particularly original student question shows how an individual tried to envision a practical situation using knowledge gained in his anatomy and physiology course:

Question: You walk into a room and notice an old woman lying on the floor. Upon close examination, you find she has a broken hip. What is the most correct series of events that led up to the old woman being on the floor and the possible cause?

a. Woman broke her hip and fell—hypocalcemia

b. Woman broke her hip and fell—osteoporosis

c. Woman fell and broke her hip—osteoporosis

d. Woman fell and broke her hip—hypocalcemia

The student perceived the correct answer to be (b), believing that osteoporosis could cause spontaneous breakage of bones—that is, that the breaking of the woman's hip preceded the fall and was not caused by it. Neunteufel says this possibility had never occurred to her before, and led her to clarify for the class how brittle bones would break in a case of osteoporosis. In fact, (c) is the correct answer.

Although Neunteufel did not use the question on an examination, it alerted her to students' misconceptions about osteoporosis and to her need to explain more clearly the difference between osteoporosis and hypocalcemia.

"QUASI-TAKE-HOME" EXAM, POINTS AWARDED FOR ASKING QUESTIONS ON E-MAIL, "EQUAL OPPORTUNITY" FOR DIFFERENT LEARNING STYLES ON EXAM

C. Walter Ogston, Department of Biology, Kalamazoo College, 1200 Academy Street, Kalamazoo, MI 49006; TEL: (616) 337-7010; FAX: (616) 337-7251; E-MAIL: ogston@kzoo.edu.

Courses Taught

- Cell Biology
- Microbiology
- Genetics

Description of Examination Innovation

Walter Ogston gives his biology students what he calls "quasi-take-home tests." The tests, which are a combination of essay, short-answer, and other formats, are given out on Friday, and students can study and discuss the exams over the weekend. They can even ask their instructor specific questions using e-mail. The instructor broadcasts both questions and answers to the whole class.

At the next class meeting, students write answers to the questions in class, without notes or books.

Unlike open-book take-home exams, Ogston says this technique helps students conceive and compose a logically coherent argument. It encourages them to think about essential points and not get bogged down in unnecessary detail. Ogston says his students view this kind of exam as a "learning experience rather than an ordeal." Moreover, Ogston has noticed that because students often send him questions (via e-mail) while they are reading or going over their notes in class, their e-mail questions tend to be more reflective than those asked in class.

Ogston also uses e-mail as a medium for submitting and responding to homework papers, and for discussion and sending notices to his class. He set up a "system alias" for mail on the Unix operating system, and all mail sent to this alias is distributed to the whole class and the instructor. To make the system support active discussion, the instructor rewards students (with a small number of grade points) for sending at least one message a week.

In addition, for the last 2 years, Ogston has been using exams that give "equal opportunity" to the different learning styles of his students. Each test has three to five sections covering different conceptual areas. Within each section, he gives students a choice of (usually) three questions to answer, each aimed at a different style of learning or expression such as discussion, construction of a hypothesis, and definition with supporting examples. In one question, for instance, students must fill in phenotypes in a table and explain how they deduced the phenotype; in another, students describe the major pathways of DNA repair and discuss long-term evolutionary consequences for different environmental scenarios. Each student must choose one question to answer from each section.

Ogston reports that few of his colleagues have used similar assessment methods in their upper-level biology classes.

PROPOSING ORIGINAL HYPOTHESES: TAKE-HOME, OPEN-BOOK GROUP WORK

Roc Ordman, Biology Department, Beloit College, 700 College Street, Beloit, WI 53511; TEL: (608) 363-2286; FAX: (608) 363-2052; E-MAIL: ordman@beloit.edu.

Courses Taught

- Biochemistry
- Microbiology
- Molecular Biology

Description of Examination Innovation

To encourage his students' research skills, Roc Ordman gives quizzes asking students to propose an original hypothesis and experimental methods to test it.

"The question is relevant in any course that addresses research methods," says Ordman, who uses it with freshmen through senior students in his biochemistry, microbiology, and molecular biology courses. A typical question is based on the science they've covered in class and readings, and is a take-home, open-book group exercise. He also gives the question individually but prefers collaborative work because the students learn more by talking with each other. And, they usually end up with more interesting ideas, says Ordman.

For example, students in biochemistry are asked to suggest and explore a hypothesis about the nature of macromolecules. The instructions remind students to "ensure that your statement is a hypothesis," and that the answer is not obvious from class reading or lecture material. Students must also explain why they feel their hypothesis is interesting and significant. In higher-level classes, students must suggest more specific research hypotheses, often based on readings for class, and must propose in some detail the methods they would use to test their hypotheses. Or they are asked to hypothesize that they are members of a review panel considering a selection of research proposals for funding. They must list the criteria they would use to evaluate these proposals and then rank them.

The question is worth approximately 30 percent of the students' quiz grade, and since Ordman doesn't give many tests, the questions can end up representing 30 percent of the final grade. Usually, students have a weekend to complete the quiz, which includes other questions asking for explanations, such as how to apply research methods to novel situations.

Students' answers are limited to one page. This way, the instructor feels he's also teaching students to be concise. The challenge is to incorporate only that technical detail into their answers that demonstrates a grasp of the concept in the hypothesis and the practical application of the methods: "I'm going to figure out how to cure AIDS," for example, is not an acceptable start to an answer, though until they begin to confront practical experiments and results, many students supply such responses. Students must select specific, workable hypotheses and explain how they would test them. Thus, they learn how to select a small and important aspect of a problem and come up with an insight that can be tested—for example, after 2 weeks of studying bacterial growth, hypothesizing at what temperature a particular bacterium will grow fastest.

In the beginning, students are unclear about what an experimental hypothesis is. As a result, they are nervous the first time they encounter this type of question. "We spend a lot of time talking about the nature of a hypothesis," says Ordman. To give them practice, he provides a no-penalty quiz within the first 2 weeks of the semester and thereafter goes over the student hypotheses and methods, discussing what makes a hypothesis fruitful. This allays some initial frustration and confusion, and leads students to develop better hypotheses on exams.

Students discover more than the power of the scientific method through this kind of testing. One student in biochemistry hypothesized that the different causes of diabetes could be distinguished using a Scatchard plot. "Five years later, the method was picked up by the medical community," Ordman says, adding that other students have come up with "prize-winning" hypotheses.

The grades Ordman gives for this exercise are simple: plus, check, and minus. A plus is awarded for an insightful hypothesis, whereas a check is for an acceptable hypothesis and method that lacks additional creativity or insight. A minus is given when a hypothesis is unclear or the method of verification is unworkable.

Quiz questions like these contribute to exciting discussion and also help the instructor identify those who haven't yet found their "creative spark." With 20 or fewer students in his class, Ordman has the time to provide extra help to these students so that they, too, gain "real-world" research experience and skills. Such encouragement to explore their own ideas may contribute to the fact that a large proportion of Beloit's biochemistry majors pursue careers in science or education, and nearly 90 percent pursue advanced degrees.

GROUP TAKE-HOME EXAMS INVOLVING DATA INTERPRETATION

Len Reitsma, Boyd Department Natural Sciences, Plymouth State College, Plymouth, NH 03264; TEL: (603) 535-2558; FAX: (603) 535-2723; E-MAIL: leonr@psc.plymouth.edu.

Courses Taught

- General Biology
- Vertebrate Zoology
- Ecology
- Ornithology
- Animal Behavior
- Tropical Biology
- Conservation

Description of Examination Innovation

In his advanced ornithology class of only six students, Len Reitsma was giving standard essay and matching exams throughout the semester. But halfway through, as a lark, he handed his students nine pages of graphs and tables and gave them an altogether different assignment: to get together outside of class, discuss the data (in most cases from other texts, but without explanatory legends), and then individually write an explanation of the data. The explanation had to include an accurate interpretation of the axes on the figures (or of the rows and columns on the tables) and a concise conclusion regarding the relationship between the two variables.

The students worked harder outside of class on this exam than the instructor had expected and did well on it, but more importantly, they found it "challenging," "fun," "provocative," and "worthwhile." One student even told Reitsma that he thought an exam such as this one should be mandatory for all science majors because it forced him to analyze and draw conclusions from data. These days Reitsma gives four in-class examinations (dropping the lowest exam grade), and the collective assignment is worth approximately one-fourth of the course grade.

Reitsma is presently team-teaching animal behavior with a colleague who is excited about using the "take-home" design for one of the exams in that course. Reitsma says that although he hasn't considered using this format in a large (40–50 students) general biology class (he has no TAs), it has been used repeatedly in upper-level courses.

"CONCEPTESTS"—WRITING ON SCIENCE CONCEPTS

Maria Schefter and Christopher S. Lobban, Division of Natural Sciences, University of Guam, Mangilao, GU 96923; TEL: 011-671-734-9533; FAX: 011-671-734-1299; E-MAIL: xmscheft@uog9.uog.edu; clobban@uog9.uog.edu.

Courses Taught

● Environmental Biology

Description of Examination Innovation

Maria Schefter and Christopher Lobban have modified Eric Mazur's innovation, the multiple-choice Conceptest, so that it improves students' ability to write connected sentences about science concepts.[3] Their innovation results in a Conceptext.

Students develop a Conceptext by reading a passage in their textbook and (if appropriate) connecting it to a relevant field-trip experience with their classmates. This can be done in groups or individually. Next, each student writes a short paragraph explaining how the textbook concept operates in a local situation. The writing can also be used simply for students to summarize a concept in their own words.

Below is an example of an effective Conceptext, which was awarded 5 out of 5 possible points. Note that grammatical errors do not impede the instructors' understanding of the content and thus were ignored in the grading:

> Laterization is the process when soil is exposed to air and the soil dries out, it hardens making it difficult for plants and roots to grow. The process begins when a forest is deforested with fire or by humans crises to cut down tree. Savanna replaces the deforested areas with grass and shrubs. But the savanna is destroyed with cars and trucks that overruns the area making it very hard to replace the area because the soil is pressed to the ground. This soil hardens and forms into cement causing laterization.

A poor Conceptext would probably include incorrect science, irrelevant statements, and be longer and more rambling than this one.

The instructors have found that students do not automatically organize material well. By rhetorical analysis of student texts, Schefter defined a successful Conceptext. It is accurate, relevant, and logically ordered. It shows a clear connection between examples and the concept being illustrated and uses clear discourse structure. Conceptexts can be used to help focus students'

---

[3] Eric Mazur's innovation is described in "Students Teaching Students: Harvard Revisited," *Revitalizing Undergraduate Science* (Tucson, AZ: Research Corporation, 1992), pp. 114–122. See also *Peer Instruction: A User's Manual* (Upper Saddle River, NJ: Prentice Hall, 1996).

attention on cause-and-effect relations, as well as drawing their misconceptions to the instructors' attention. Conceptexts are very much like short essay questions, common on tests, so that instruction with the diagram helps students hone critical-thinking skills that are useful in making the best use of examination time.

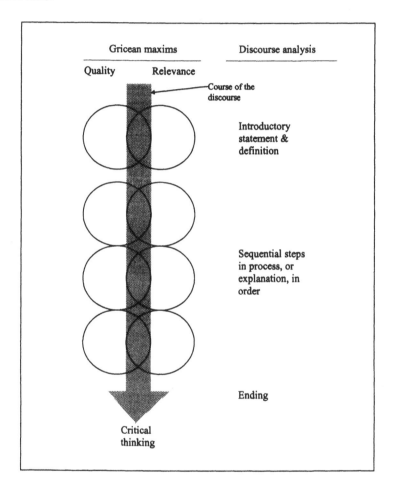

According to the instructors, students appreciate gaining insight into how to write answers to essay questions.

Lobban recently started teaching nonmajors after years of lecturing to upper-division majors. With Schefter, who has an education/linguistics background, the two instructors have been working together exploring the language and culture of science, including coauthoring *Successful Lab Reports*, a book on writing lab reports (New York, Cambridge University Press, 1992).

Schefter and Lobban have not yet shared this technique with their department colleagues, but they intend to do so with government funding they have received to build bridges between English and biology teaching.

GROUP QUESTIONS ON QUIZZES

Patrick Thorpe, Department of Biology, Grand Valley State University, Allendale, MI 49401; TEL: (616) 895-2470; FAX: (616) 895-3412; E-MAIL: thorpep@gvsu.Edu.

Courses Taught

- Introductory Biology II
- Human Genetics
- Cell and Molecular Biology (and lab)
- Seminar on Evolution (capstone)

Description of Examination Innovation

Although Patrick Thorpe surveys and then places his students into study groups balanced by gender, GPA, and coursework, he finds this does not guarantee they'll work together—which he wants them to do. As incentive, for several years he has been giving group questions on quizzes, in which one member of the group is responsible for the grade awarded to all group members.

Significantly, the students do not know who will be chosen to answer the group question, so they are encouraged to meet to make sure all group members are equally prepared. Thorpe chooses students randomly, rotating throughout the term. The group question is usually based on a topic the group was supposed to study in preparation for the quiz. For example, in cell and molecular biology, a group question asks students to answer two analytical questions about a 1938 high-school textbook discussion of normal vs. feeble-mindedness as genetic traits. The rest of the quiz is completed individually.

"What I find is that when students get together to make sure everyone knows the answer, they wind up studying together," says Thorpe.

Thorpe recently expanded group work in his cell and molecular biology lecture classes. Groups draft detailed outlines for the particular lecture to which they are randomly assigned (Thorpe provides them with short lecture outlines). The group turns in the detailed outlines, and Thorpe edits and returns them. Someone in the group then retypes the outline and distributes it to the class. Thorpe grades the outlines only on whether they are completed, but because exams and quizzes are written from these outlines, students come to realize that a poorly constructed outline will make it harder to study for the test. In the best cases, all students discuss the outline to ensure its quality and usefulness for studying.

In human genetics, Thorpe has groups work problems on the blackboard and post answers for the rest of the class. He has tried using groups in introductory biology, but with 95 students, he had too many groups to deal with. He also found these introductory students more competitive—the "Why should I help anyone else?" mode, he calls it—than his higher level students.

DRAWING QUESTIONS ON EXAMS

Lucia Tranel, Department of Biological Sciences, St. Louis College of Pharmacy, 4588 Parkview Place, St. Louis, MO 63110; TEL: (314) 367-8700; FAX: (314) 367-2784; E-MAIL: ltranel@slcop.stlcop.edu.

Courses Taught

- Anatomy
- Physiology
- Biology

Description of Examination Innovation

Anatomy is a visual science, but in traditional anatomy/physiology classes, many students study the material by memorizing pictorial representations, performing dissections, examining models, and labeling drawings. They do not necessarily learn to relate form to function, says Lucia Tranel, unless they can retain detailed images of micro- and gross-anatomical structures, which is why, for the past three years, Tranel has had her students do a lot of drawing—in class, for homework, and in exams.

"If students are looking at someone else's picture to learn anatomy and physiology, it gives them a visual reference for what they're trying to learn, but it's not the same as if they produced the drawing themselves. It's similar to looking at a map and trying to memorize how to get somewhere versus drawing the map yourself," says Tranel.

"Rarely is anatomy taught actively to students," says Tranel. Rather, it is assumed that the anatomical information will be absorbed automatically, along with the physiology of bodily processes. Tranel has learned not to rely on osmosis. She has her students diagram, label, and describe human anatomy.

In her lectures, she does drawings in front of the students, using transparencies, and provides students with a lecture workbook that includes blank pages, as well as preprepared drawings of many of the structures she covers. Tranel encourages her students to draw along with her and to take written notes right on the diagram.

"This new approach has so changed how I view student learning that I now call my course, 'The *Art* of Human Anatomy,'" she says.

Tranel's novel technique was the result of an "accident," she says. After a frustrating attempt to get one of her students to draw, from memory, a picture of a nephron, she finally asked, "Could you draw a picture of your dorm room?"

"Yes," said the student.

"Why?"

"Because I know what my room looks like. I can see it in my head."

Tranel decided she would try to help her students *see anatomy in their heads.* Drawing turns out to have many additional, unexpected advantages. It accommodates visual and other nontraditional learning styles, measures precise information without employing language, and permits the less verbal student to demonstrate what he/she knows. When students are required to draw from observations made in the laboratory, Tranel can immediately determine if the student is "on the right track." The drawing Tranel requires is not dependent upon artistic ability; it is easier, she insists, for a nonartistic student to do an anatomically correct line drawing than for a language-impaired student to write a short essay under exam conditions. Finally, diagrams are easier (less susceptible to dispute) and less time-consuming for the instructor to evaluate than written exams.

Tranel also uses diagramming in examinations, which benefits the students because an instantaneous self-assessment is possible, with immediate feedback about misconceptions. No one, Tranel points out, can draw the details of a structure he/she can't visualize, and therefore hasn't "mastered." Tranel says she puts "a lot of thought into each question" before including it, so as to avoid ambiguity, and she usually requires a written description along with the diagram.

"Usually, my drawing questions are very specific, and I might even tell students to be sure to include certain components in their drawing. But grading is still a challenge," she says. "I try to be objective: If I can see that the concept is there—things are on the inside that belong on the inside, or the student has put layers in proper order—then I accept the drawing. 'Drawing to learn' is a process that benefits us both."

Students do appear to be learning better. In a small study comparing two classes (one at a community college, one at St. Louis College of Pharmacy), students seem to retain longer more of the finer anatomical details after being made to draw in preparation for an examination, most likely because they build a reference framework that appears to last beyond the course, according to Tranel.

COMPUTERIZED ASSESSMENT

Frederick R. Troeh, Department of Agronomy, Iowa State University, Ames, IA 50011; TEL: (515) 294-3273; FAX: (515) 294-8146; E-MAIL: frtroeh@iastate.edu.

Courses Taught

- Soil Fertility

- Soil Resource Conservation
- Soil, Fertilizer, and Water Management
- Agronomy 211 (a career opportunities seminar)

Description of Examination Innovation

Students in the soil fertility course taught by Frederick Troeh and Thomas E. Loynachan now do their term project on computers rather than on paper. The four-part soil-sampling problem is graded almost entirely by a computer program developed by Troeh, with assistance provided through instructional improvement funds from the university. The problem has been through several revisions since its introduction in paper form many years ago. The computerized form, first developed in 1990, is now an effective and greatly streamlined assessment tool.

Troeh says that in a course such as soil fertility, posing real-world problems is the best way to teach as well as evaluate students. For students to demonstrate that they've learned how to manage soil fertility, they must use decision-making skills and calculate the results of their decisions in specific applications. The computer allows students to make that demonstration as realistic and thought-provoking as possible. Troeh's project is worth one-sixth of the students' final grade and takes about 8 hours of work spread through 12 weeks of each term.

Each student receives a computer disk with a different version of the problem to solve. Students are allowed 2 weeks to solve each of the four parts, with 1 week in between for recording grades. Students may use the computer lab or their own home computers to work on the problem outside of class. The program runs on an IBM-compatible computer, preferably with a color monitor.

Part I begins with a soil map (from various Iowa counties) on which students must indicate where they need to take samples as a basis for fertilizer recommendations. After marking at least 15 subsample locations on the map for each of several (up to nine) soil samples, they must designate what soil tests are to be performed on each sample. Students complete Part I by calculating the costs and answering some questions about how costs may be justified.

Part II of the problem provides the student with a set of soils and the results from the soil tests performed on their samples. The student must obtain additional information from references about the location of the soils in the state, the climate of the area, and the significant physical characteristics of the soils, and then make recommendations for how much of each fertilizer nutrient and/or lime is needed to grow specified crops.

Part III deals with adjustments the producer should make in recommendations that come from the soil-testing laboratory and adjustments to recommendations for each new yield goal. Possibilities include the growth of a legume that provides nitrogen for the next crop or a recent application of manure or lime that has not yet fully reacted with the soil. Students must also know how to adjust the recommendations for each new yield goal. Part IV involves choosing appropriate fertilizers and calculating the costs and anticipated profits according to three different approaches.

An advantage to this project, says Troeh, is that students can help one another through the problem—in getting used to its broad outlines—but ultimately, they are forced to work individually, since their respective versions of the problem have different numerical parameters and are based on different soil maps.

Almost all (90–95 percent) of the students, according to Troeh, prefer the computerized version of the problem to the paper version. (A small percentage indicate on end-of-course evaluations that they didn't like working on computers.) Troeh feels they're learning more about soil fertility and soil tests and seem to be more interested in the problem when they can use computers to solve it.

For Troeh and his staff, computer-assisted project assignments make evaluation of student work more meaningful both to the student and the instructor. First, the computer works in tandem with the student, providing immediate feedback. Troeh explains that after a student answers one section, the computer asks if he/she is satisfied with the answers. Then it grades the answers, recording the grade on the diskette, and if errors were made, the student is immediately instructed in the correct answers. When the paper version was used, grading each section took 1–2 weeks. Feedback was delayed and, thus, lost its impact. Uncorrected mistakes made in the early part of the project sometimes led to highly improbable results and conclusions.

Second, the computer takes over much of the grading for Troeh, permitting him to concentrate on the students' approaches to soil sampling. Part I, in which the student maps out his/her approach to sampling, must be graded by the instructor and graders. (A grading program assists by having the computer display all of the student's work on a summary screen.) But Parts II, III, and IV, which contain numerical calculations and multiple-choice questions, are graded by the computer without supervision. Finally, the program reads all the student files and tabulates the results.

"Not only do we get a grade for each student," says Troeh, "but also the program identifies class performance on various parts of the problem, so we can tell which parts are causing the most difficulty."

GROUP EXAMS

David B. Wagner, Department of Forestry, University of Kentucky, Lexington, KY 40546-0073; TEL: (606) 257-3773; FAX: (606) 323-1031; E-MAIL: for114@ukcc.uky.edu.

Courses Taught

- Introduction to Population Genetics
- Issues in Agriculture: The Development of Modern Agriculture

Description of Examination Innovation

In nearly two decades of university teaching, David Wagner has never before witnessed the high level of energy and "vigorous discussion" that was apparent in his classroom in the spring 1994 semester, when he used group exams for the first time in his population genetics course.

The midterm included two quantitative group problems for which each group of six students submitted answers. The remaining 80 percent of the midterm consisted of individual questions. Students worked on their midterm in the same study groups to which they had been assigned by Wagner at the beginning of the semester.

Wagner decided to try group exams because he thought that by being forced to agree on a single answer to an exam question, students would learn better. He was so impressed by students' performance on the midterm that for the final exam, he assigned two group take-home quantitative problems (with subparts) worth 50 percent of the exam. The remaining 50 percent of the final was based on an individual take-home, open-book question. Students were encouraged by Wagner to discuss ideas with each other, but each student submitted an individual written answer. Wagner observed that discussion between students for this individual portion of the final "seemed to benefit the understanding of weaker students. Yet for grading purposes, it remained apparent which students really 'got it.'"

In addition, Wagner allotted 15 percent of each student's final grade to peer evaluation by fellow group members of that student's individual contribution to the group's activities, an idea he borrowed from a colleague in the Department of Sociology at the University of Kentucky. Hardworking students in Wagner's class found this to be a fair and effective way to penalize any "freeloaders" in the groups.

Wagner sees the active engagement of students in discussion as an additional benefit, but in conversation and in course evaluations, he finds his students do not (yet) think these benefits justify the extra work required of them. His colleagues have responded with both interest and skepticism to the group format for examinations.

GRADED LABORATORY EXERCISE

Richard Walker, Department of Biology, Sterling College, Sterling, KS 67579; TEL: (316) 278-2173; FAX: (316) 278-3188; E-MAIL: dwalker@acc.stercolks.edu.

Courses Taught

- Genetics
- Human Heredity

Description of Examination Innovation

Ten years ago, Richard Walker decided to employ a lab test in his genetics course in place of the usual 50-minute exam.

"Genetics is a peculiar beast in terms of testing," he explains. "If you really want to see that students can take complex data and analytically break it down, explain it, and test a hypothesis, 50 minutes is a difficult time constraint. You can do it, but you can't ask sophisticated questions." This is why Walker developed a genetic analysis exercise.

The laboratory exercise involves a complex analysis of "gribbles," an imaginary species he describes as "peculiar animals found only in the far reaches of the tundra near the Arctic Circle," with bodies that, when preserved, "strongly resemble styrofoam." The gribble is actually a piece of styrofoam packaging. With the variety of shapes and colors available, sexual dimorphism and different phenotypes, such as stripes and dots, are easy to produce. Two "discoveries" are detailed in the instructions to the exercise: two distinct and heretofore unseen abdominal markings on gribbles in two different regions.

Students are assigned to groups of three, whose task is to apply the given information to determine the inheritance patterns for abdominal markings in the two gribble populations. To do this correctly, they must construct a hypothetical model using chi-square analysis to explain the data and to prepare a written report. The inheritance patterns included are dominant–recessive, epistasis, X-linkage, and lethal genes. Students are given 2 weeks to complete the assignment outside of class. They may use any resources they wish. Although technically a laboratory exercise, this assignment counts as the equivalent of half a major exam grade.

At first, students resist exercises such as this because they are harder and not the kind of exam questions they can look up in books. "They have to stop and think, and use what they know," Walker says. "But by the time they finish the course, they're doing well."

OPEN-BOOK EXAMS USING REAL DATA

Angela Wandinger-Ness, Northwestern University, Department of Biochemistry, Molecular Biology, and Cell Biology, 2153 Sheridan Road, Evanston, IL 60208-3500; TEL: (847) 467-1173; FAX: (847) 491-2467;
E-MAIL: w-ness@nwu.edu.

Courses Taught

- Cell Biology

Description of Examination Innovation

Angela Wandinger-Ness uses recent publications in scientific journals such as the *Journal of Cell Biology* for exam questions in cell biology. She simplifies data, in the form of figures or tables, and creates exam questions that require students to apply their factual knowledge to interpret the results and draw conclusions from the data.

Because all of Wandinger-Ness's exams are open-book, the questions replicate the challenging nature of research. Graph, data, or gel results from a paper are selected and reproduced with information about how the experiment was performed. Questions on recent cell biology midterms explore the posttranslational modification of microtubules as measured by the incorporation of radiolabel into tubulin, protein translocation studied by monitoring the function of proteoliposomes reconstituted from purified endoplasmic reticulum proteins, and the regulation of cell-cycle-dependent kinases by monitoring their phosphorylation and dephosphorylation.

One disadvantage of this method may be encountered during take-home exams. Wandinger-Ness has had students actually find and use the original reference for her exam. Although this shows initiative, it certainly affects the reliability of the method for assessment and is why other faculty at Northwestern use this method strictly for in-class assessment.

Although some students complain that the grading (including partial credit) isn't "fair"—meaning written explanations aren't as "objective" as numerical answers—most of the instructor's students appear to prefer biology questions for which they must think rather than memorize.

LAB GROUP QUIZZES IN MULTIPLE FORMATS

Margaret A. Weck, Saint Louis College of Pharmacy, 4588 Parkview Place, St. Louis, MO 63110; TEL: (314) 367-8700 Ext. 1305; FAX: (314) 367-2784; E-MAIL: mweckm@slcop.stlcop.edu.

Courses Taught

- Human Anatomy and Physiology
- Human Physiology

Description of Examination Innovation

About 3 years ago, Margaret Weck began using group quizzes in her anatomy and physiology laboratories. In addition to having the support for collaborative learning strategies from her school, Weck recognized the lab as an especially good place to experiment with these techniques. After all, labs contain smaller numbers of students than lectures, and the students are already naturally organized into working groups. Finally, Weck wanted to make the traditional, rather mediocre lab obsolete—the one in which students focus on a manipulation without understanding why they're doing so.

In Weck's group quizzes, groups of three to four students work together to answer factual recall, analysis, and application problems on the material studied in the lab. The format for her group quizzes varies with the content being assessed. Questions may ask students to label muscles in a cat dissection or name certain bone landmarks on a model; others cover how to calculate the number of blood cells or perform other laboratory techniques, especially if those techniques and procedures require students to understand the physiology or chemistry underlying them, as in doing a urinalysis. Weck says modified multiple-choice questions, in which students can choose more than one answer, and modified matching questions, in which there are more items in one column than in the other, work to stimulate group discussion almost as well as short-answer questions, because all these formats require students to defend their choices to the group.

Students may also take similar quizzes individually on different occasions. In any given week, they will not know in advance which kind of quiz they will have, so they have to be prepared for both. "Sometimes students do the least amount of work to get a grade," says Weck. "So, I deliberately do not tell them which type of quiz to prepare for. I want to enforce the idea that they're all individually responsible for their understanding."

More recently, Weck has started giving two quizzes during each lab. The first is individual and comes at the beginning of the period. Usually, it covers what was on the previous week's laboratory and underscores the need for each student to be individually accountable for understanding the procedure and reason for doing the lab—for getting the big picture, so to speak. The other, a group quiz, is done at the end of the lab. The group quiz "consolidates students' understanding of what they just did together, often under time constraints," says Weck, "and sets them up for putting the experience into context and understanding it better over the week." Group and individual quizzes are weighted equally. Weck's practical exams, for which the group quizzes prepare students, are individual and worth more than quizzes.

Weck admits that she has never been entirely comfortable giving one grade to all members of a group. She thinks faculty are not socialized to feel comfortable assessing group dynamics, so it's hard for her to use a group dynamic component in their science grade. To compensate for the confounding aspects of collaborative work and assessment, she has built in what she calls "safety valves" for students who want to make sure that the work for a grade is fairly distributed. First, she gives the group the option of excluding anyone who doesn't contribute at all; however, no group has yet used this extreme sanction. Although it is extremely difficult for students, especially the freshmen in her class, to confront one another about problems in the group, Weck feels this option keeps students from sloughing off and provides a preview of the kind of teamwork that industry and many professions will expect of them later on.

There are always some "dysfunctional groups" that can't resolve their conflicts. Therefore, Weck offers as many safety valves as possible. Those not good at speaking up have a way to make their opinion count, and the very outspoken can be curbed.

Group members draw up "rules" for the group and grade each other's performance (participation points) throughout the semester in light of their published rules. Each member gets a copy of his/her group's rules, and the instructor keeps a copy. The instructor specifically asks groups to draw up rules addressing how they will define participation, how they're going to make up missed labs, and how they are going to deal with or settle disputes.

"These methods have never lowered anyone's course grade," she says, "but the students feel good about having this power."

Indeed, Weck says the amount and quality of interaction in class increases dramatically when students realize they will be graded by group members on their contributions and comprehension.

She notices more group discussion and group teaching. So long as checks and balances exist, Weck is comfortable assigning a single, shared grade for each assignment completed by the group.

A recent evaluation of the lab component of the course indicated that students find the group process helpful and enjoyable. Weck specifically asked a colleague to perform a small-group instructional diagnosis at the beginning of the spring 1994 semester, just prior to instituting the change to the group quiz format. Two questions were asked: "What has helped your learning in this course?" and "What changes would further help?" Students said they feel the group quizzes help them learn the material, either by explaining their reasoning to their peers or learning from someone else who understands it better and can explain it. Students say they feel less pressured taking group rather than individual quizzes, yet Weck's comparison of group and individual quiz taking reveals no significant difference in average scores using the two methods.

The transition to group work is hard for some students, giving rise to complaints that the teacher isn't teaching, the students have to do too much work, and it isn't clear how to tell what's most important in all of what they're doing. Weck feels that some of this frustration is a necessary experience for students going on into the world of work.

EXAMS BASED ON CLASS GOALS: TEAMWORK, DATA ANALYSIS, PROJECT DESIGN, COMMUNICATION, AND PEER-REVIEW SKILLS

Robert Yuan, Department of Microbiology, University of Maryland, College Park, MD 20742-4451; TEL: (301) 405-5436; FAX: (301) 314-9491; E-MAIL: RY11@umail.umd.edu.

Courses Taught

- Microbes and Society
- Biology for Nonscience Majors

- Principles of Biology and General Microbiology

Description of Examination Innovation

In recent years, Robert Yuan and Spencer Benson have revised their courses to reflect skills required not only in biological research but also in a working environment. The skills include teamwork, data analysis, project design, writing (in upper-division courses, research proposals), oral presentation, and peer-review and electronic communications. The goal of this work, in the words of the instructors, is "to train students in critical thinking and scientific skills, and expose them to the environment of the workplace."

Each of these skills, separately and together, is incorporated into the class's team-testing strategies. To advance teamwork, groups of four students work together in class on pop problems and quizzes. In laboratory, teams do problem assignments and conduct student-formulated laboratory experiments "in concept"—by preparing their own protocols and developing and conducting their final class project. And in the advanced microbiology course, each four-person team reads and discusses reviews of the literature and collaboratively prepares research proposals.

The instructors incorporate data analysis into the problem assignments and tests given in all of the upper-level courses, as well as into the laboratory component of these courses. Writing assignments vary by course and include minor and major essays allowing rewrites, laboratory reports, research proposals, and a written evaluation of a business plan by a new biotechnology company. Oral presentations are done by individuals within groups in the form of "full-dress presentations" involving the use of overhead projections, time limits, and questions from the audience and the instructor.

But the peer review component is the most radical of the new testing procedures. Recognizing that an important aspect of scientific work involves the critical review of the work of colleagues, as well as accepting constructive criticism as the basis for improving one's own work, in the upper-division courses, oral presentations and research proposals are subjected to peer review. Each proposal is given a group grade. But then each member of the group evaluates the other group members according to effort and productivity. These peer evaluations are then used to convert the group grade to individual grades.

The separate grading of course assignments (e.g., research proposals, laboratory projects, and oral presentations) allows the team to monitor the acquisition of specific skills and to note progress made during the semester. Students appear to perform better under the revised testing strategies, and the instructors get a "valuable window" into the functioning of individual students.

CHAPTER 3

CHEMISTRY

"CAPITALIST" GRADING SYSTEM

Philip Beauchamp, Chemistry Department, Cal Poly–Pomona, Pomona, CA 91768; TEL: (909) 869-3659; FAX: (909) 869-4396.

Courses Taught

- Organic Chemistry

Description of Examination Innovation

Phil Beauchamp has introduced a "capitalistic reward system," as he calls it, in his innovative method of grading his organic chemistry courses. Up to 30 percent of the grade can be earned by doing extra problem sets of the sort he would normally feel constrained to give students, since they are difficult (some of them he calls "massive"), time-consuming, multistep problems. This means the in-class examinations can count as much as 100 percent of course grade or as little as 70 percent, depending on the students' willingness to do the extra problems. Extra credit in Beauchamp's scheme thus becomes "substitute credit." Regular homework problems are promptly graded and returned along with an answer key.

What kinds of problems are assigned for extra credit? Some require students to describe in detail the mechanisms of a reaction. Others require them to explain an observation in terms of the logic of organic chemistry. Still others require that they do or show that they understand synthetic transformations.

Also of note is the fact that Beauchamp invites his students to do the extra problem sets in groups, which particularly benefits his weaker students. Students have commented on course evaluations that "the best thing I did was to work in a group." Since scores are entirely individual, students form groups on their own.

Homework grading is done in a manner that cannot depress the student's average in the course. An arbitrary total of 30 points is given to each large homework set. If all the homework is attempted, a value of 24–30 points is given (80–100 percent). A 100 percent score is given if selected problems are mostly correct, and an 80 percent score if an honest effort was made but errors are present. If some of the homework is missing, a quick judgment is made that two-thirds, or one-third, or none of the homework credit is due, and 16–20 points, or 8–10 points, or no points are given, respectively. Any homework points not assigned are replaced by the exam scores.

During the term, miscellaneous homework assignments are assigned, as desired, and added onto the homework point total. These are usually counted as 10 points each and graded 10, 9, 8, or no points. If all the homework is done, it will represent 30 percent of the grade, and exams will count 70 percent. If half of the homework is done, it will contribute 15 percent of the grade, and the exams will count 85 percent, and so forth, on a sliding scale. If none of the homework is done, then the exams will contribute 100 percent of the grade.

Beauchamp allows students to substitute the comprehensive final exam for the entire course grade if the final exam grade is better than their overall course grade. This keeps students from dropping the course, since they can salvage their course grade on the final exam. All questions on the final require active participation on the student's part, as there are no multiple-choice questions. Beauchamp estimates that 5 percent of the students obtain a higher grade in the course than their course average as a result of their performance on the final exam.

To reduce exam anxiety, Beauchamp also allows students to bring in one $8\frac{1}{2}$" × 11" page of notes. The notes appear to have little effect on exam performance but somewhat reduce exam anxiety. Students spend considerable time and effort organizing and making up these note pages, which, of course, organizes their thinking and recall as well.

The only problem for the instructor is that it takes a few hours longer to calculate final grades at the end of a quarter. Each student could have any possible homework contribution between 0 and 30 percent, and an exam contribution of 100 percent minus the homework percentage. An individual calculation is necessary for every student. This can be done quickly, with practice, in a few key punches on a calculator.

Beauchamp is deeply committed to his grading system, and so are his students. "I've never had a student complain about the system," reports Beauchamp, who has used it for over 10 years. Fewer than 5 percent of the students take the 0 percent option—that is, do no extra-credit problems. Most do as many as they can. "It is a great motivator. I get students to do difficult homework they would otherwise resent," Beauchamp says. Since he doesn't have to force any of his students to engage in the extra work, he sees this as a "capitalist" option. He is "paying" students for their effort. And students understand this. In final course evaluations, students almost unanimously take responsibility for their performance.

In response to the question, "Are students doing better?" Beauchamp says those in the middle range are doing better than they would have—better in both senses: getting a better grade and learning more because they are working on harder and more interesting problems.[1] In any case, fewer students drop out of the course.

Still, enthusiastic as Beauchamp is for his system, he reports that no one else in his department has adopted it. Why? "Too much work," he says.

INDICATING POINTS LOST DUE TO NOT USING DIMENSIONAL ANALYSIS

Henri A. Brittain, formerly of Kennesaw State College, Technology Leader, Control and Information Systems Development, The Procter and Gamble Company, Winton Hill Technical Center, 6300 Center Hill Avenue, Cincinnati, OH 45224-1795; TEL: (513) 634-3397; FAX: (513) 634-2439; E-MAIL: brittain.ha@pg.com.

Courses Taught

- Physics (trigonometry-based)
- General Physics (calculus-based)
- Medical Physics (special topics course)
- Industrial Chemistry (special topics course)
- Biostatistics
- Fundamentals of Physical Chemistry (for biology majors)

Description of Examination Innovation

With a background in chemical engineering, Henri Brittain believes that mastery of dimensional analysis is important in problem solving. Last year, when he taught at the department of chemistry and biophysics at Kennesaw State College in Marietta, Georgia, between 30 and 40 percent of Brittain's students had difficulty with dimensional analysis, despite constant reminders to "check your units." Although he works through dimensional analysis at the chalkboard for every problem, some students either don't care or don't understand enough to do the same on their exams.

As a new tactic, Brittain now assesses the number of points lost on an exam that could have been avoided had dimensional analysis been employed and reports this back to every student. Such errors include using the wrong equation, failing to change units such as rotations per minute to radians per second, omitting values for variables once the equation has been written out, errors in

---

[1] Upon request, Phil Beauchamp will make available his grade point averages calculated by gender.

carrying out algebra, and so forth. The process is simple for Brittain and revealing for his students. The instructor himself has been surprised to discover that fully half a letter grade is often lost because of mistakes such as these. His students have been surprised as well.

Many students have told the instructor they were surprised to learn that familiar equations had units, having used these equations for years in math courses in which equations are dimensionless expressions. Students are also beginning to believe the instructor when he tells them that they can reconstruct a forgotten equation through dimensional analysis. When one of Brittain's former physics students couldn't remember, during her MCAT exam, that the product of wavelength and frequency equals the speed of light, she was able to figure it out by recalling the units of each variable.

TAILOR-MADE TEST-ITEM BANK, LAB PRACTICAL

Joseph Casanova, Department of Chemistry and Biochemistry, California State University, 5151 State University Drive, Los Angeles, CA 90032-8202; TEL: (213) 343-2338; FAX: (213) 343-6490; E-MAIL: jcasano@csula.edu; ChemItem WEB SITE: http://www.cchem.berkeley.edu.8080/.

Courses Taught

- Microcomputers in Chemistry
- Organic Chemistry Lecture and Laboratory
- Advanced Organic Chemistry and Laboratory
- Polymer Chemistry
- Chemistry of Secondary Plant Metabolism

Description of Examination Innovation

In an effort to ease the clerical burden of examination preparation, in 1985, Joseph Casanova and his colleagues began collecting test and homework items for organic chemistry using the microcomputer as a platform, thus removing the need for keyboard entry of test items. In 1987, they sent electronic and printed copies to organic chemistry faculty on all California State University campuses. The test-item bank, called "ChemItem," has been used modestly on several campuses since.

The items in the pool have been collected personally by Casanova and his colleagues, and are drawn (usually modified) from the many textbooks used in the organic chemistry course over the years. Some, which have been tested and judged to have been effective on examinations in the past, are original items created by Casanova and his colleagues.

Data files are maintained in a manner that makes it convenient to select, assemble, and produce letter-perfect examinations and quizzes with minimal effort. (An eight-item exam can be produced in about 15 minutes). At the same time, items can be modified and tailored to a particular instructor and class, as the database is designed to be flexible enough to allow the instructor to add or revise items. Thus, although computerized, the system does not restrict the user to a uniform set of test items. Indeed, "most faculty strongly prefer to 'do their own thing' in the writing of examinations," says Casanova, "and many enjoy constructing new questions."

Casanova has continued to add to the item bank and to update the application base for its storage and presentation. Since the database has been made available on disk to be copied to a hard disk, Casanova has no way of knowing how extensively it is being used except secondhand. So far, the availability of the items has not been advertised outside the California State University system, but the bank is now on the World Wide Web (see p. 68). The availability of test items and their answers to students is not a problem, says Casanova. "The very act of studying from a large pool is a major source of learning for students," he says.

If the approach to exam preparation outlined here seems useful to readers, and they wish to help maintain and expand the item pool, Casanova and his colleagues are eager to have readers send new items in the specified format,[2] to be included in their next update.

Another innovation employed by Casanova includes the practical examination of organic chemistry laboratory skills. Concerned that the laboratory report, often prepared outside the class, represents an ambiguous measure of the quality of observation and other laboratory skills, Casanova and his colleague Linda Tunstad have designed several laboratory practicals valued at 10–15 percent of the total course grade. The exams, administered on the last day of class, require students to carry out several common laboratory procedures correctly, and to analyze the results of their experiment, discriminating among several possible initial conditions or outcomes. The instructors are aiming to test critical thinking, among other competencies, with these exams.

A typical practical challenges students to prepare and characterize one of the isomeric propyl acetates. (Either 1-propanol (bp 97 degrees C) or 2-propanol (bp 82.5 degrees C) may be supplied as an unknown C3 alcohol by the instructor.) The techniques of reaction, reflux, multiple extraction, drying, and distillation are tested. The student is supposed to deduce from the boiling point of the product (a propyl acetate) the identity of the propyl alcohol that was supplied. The alcohol

---

[2] Helvetica 12-point font in Microsoft Word, Version 5.1a on the Macintosh. Default ruler margins of 1.0 inches to be used throughout. All items to carry an expendable item serial number in the upper-right corner. Microsoft Excel is used to tabulate and display the serial number and keyword indices.

can also be identified by a micro–boiling point measurement, but only if the student understands the objective of the experiment before using all the starting material. Finally, the 1H nmr spectrum has to be sketched out and infrared cues regarding the structure of their product described.[3]

Casanova and Tunstad believe that this practical in-class laboratory final examination provides a reliable measure of students' laboratory skills with minimal time and energy from faculty and staff.

Although student anxiety levels (and glass-breakage rate) are notably higher during these exams, the instructors observe a strong correlation between performance on these examinations and their qualitative judgment (by observation) of the students' laboratory ability. Performance on these examinations correlates less well with overall performance in the course. A number of students have overall scores in the course significantly exceeding their performance on the practicum, suggesting that standard written laboratory evaluations may inadequately measure laboratory ability. The laboratory exams underscore the critical connection between experimental observation and theoretical conclusions—the very things chemistry instructors are trying to inculcate.

GROUP EXAMS, ESSAY QUESTIONS

Karl De Jesus, Department of Chemistry, Campus Box 8023, Idaho State University, Pocatello, ID 83209-8023; TEL: (208) 236-2673; FAX: (208) 236-4373; E-MAIL: DEJEKARL@fs.isu.edu.

Courses Taught

- Organic Chemistry
- Advanced Organic Chemistry

Description of Examination Innovation

For some years now, Karl De Jesus has been using some alternative testing methods, including "team problems," take-home problems, and essay questions. The "team problems" are organized as follows: Students are divided into teams of seven or less by the instructor according to their proficiency in the subject. Each team is given at least 1 week to solve its unique assigned problem and is required to present the solution orally in an outside-of-class panel discussion. Typical problems require completion of a multistep synthesis, given specific

---

[3] This experiment is available on inquiry in full-size, printed form or in BinHex 4.0 (Macintosh) format via e-mail inquiry at the address above.

starting materials; interpretation of reaction data in the identification of an unknown; interpretation of kinetic, isotopic, and stereochemical data in the elucidation of reaction mechanisms; or the interpretation of spectral data in the identification of an unknown.

During the panel discussion, the instructor poses random questions to individual members of the team. Then an overall grade is calculated as a function of all answers to the questions. Although this works well for classes of 45 or fewer students, the panel discussions are too labor intensive for larger classes. A variant that De Jesus and his colleagues will try with classes larger than 100 students is to have each team hand in a written solution and then take an in-class quiz specific to their problem. Each portion—the take-home solution and the in-class quiz—would count for half of the team problem points.

The benefit of the exercise is that it provides a mechanism for cooperative learning, at the same time allowing the instructor to truly challenge students individually. In large classes, however, De Jesus will often substitute panel discussions with a special section on the next exam for each of the different teams. This helps guarantee individual student accountability and cuts down on time. But De Jesus prefers the oral-presentation aspect of panel discussions and continues to seek "a way to penalize lack of participation without jeopardizing the spirit of the exercise."

Student reaction has been favorable. Some are intimidated initially by the new techniques but soon warm up. In fact, students often cite the panel discussion as the most useful learning tool of the course.

Another variant on team problems is to give one complex problem in chemistry (similar to a team problem) to all students in the class. Groups (no larger than eight) are allowed to work together and hand in one solution, complete with participant signatures. This practice, the instructor finds, is especially useful for students who do not test well but who are able to show creativity when the time pressure of an in-class exam is removed.

In this case, students get to decide with whom to work.

Finally, De Jesus asks essay questions on tests that require students to analyze chemical case data and draw conclusions from the data. In answering questions such as these, students are expected not only to choose the correct conclusion but also to defend that conclusion by referencing chemical principles, mechanisms, or other tools. Although the essay questions are "a great learning and training tool," the instructor finds grading them extremely laborious and, hence, does not use the essay-type question in large classes.

Three years ago, De Jesus began asking questions on exams in the following format: "Unknown A after reaction 1 gives B, which is a stereoisomer of C. Unknown C was prepared from known compound D by reaction 2. Give the structure of A, B, and C." Questions like these, De Jesus finds, force students to learn reactions forward and backward, as well as their stereochemistry. In addition, students learn to use analytical reasoning in solving problems. The problems can be as involved as the instructor desires. Sometimes De Jesus gives as a clue an intermediate structure for which students must provide the conditions.

Problems like these, De Jesus reports, are easily graded. Thus, while challenging, they provide a good balance between student effort and faculty grading time. De Jesus has used team problems at Union College, where he taught previously, and says colleagues at Idaho State have also used team and group problems successfully, though they note how much extra time these exercises require.

PROBLEMS BASED ON CHEMICAL SEQUENCES

K. Thomas Finley, Department of Chemistry, State University of New York, College at Brockport, 300 New Campus Drive, Brockport, NY 14420-2872; TEL: (716) 395-5588; FAX: (716) 395-5805.

Courses Taught

- Organic Chemistry
- Chemistry for Health Professions
- General Education Courses for Nonscientists

Description of Examination Innovation

"To the beginner, organic chemistry is like a foreign language," says Thomas Finley. Building on his analogy, and using an idea he developed with his wife, Patricia Siegel, a French professor, he has come up with some assignment and exam formats to make the necessary memorization more enjoyable and to get the students to *think* more about organic chemistry—with less focus on memorization.

Because first-semester organic chemistry students are "beginners" as far as organic chemistry is concerned, and have difficulty memorizing abstract formulae, reagents, and reactions, Finley uses a creative, helpful final examination that resembles a puzzle, with hints that are meant to make gaining a solid foundation more enjoyable. To solve the puzzle, students have to think organically about organic chemistry. Finley takes three complicated synthetic sequences, with approximately 10 steps each, and puts them on his exam with missing reagents, reactants, and products. The parts that have been "deleted" are key to understanding the chemical behavior described.

The instructor provides his students with a list of all the missing structures and asks them to place them correctly. Thus, his students don't have to worry as much about cramming chemical formulae and reactions into their brains; if they understand the material, they can relax.

"The examination becomes one of not simply knowing the chemical reactions, but also of being able to use the logic of the relationships," he says. And although use of this exam slightly raised students' scores, the exam still differentiates student abilities well enough to help Finley understand their needs.

Finley also gives his students *dictées*, in which they must listen to and then draw chemical compounds and reactions without writing down the English names themselves—without "translating" them. For example, he will "recite" a reaction out loud, first rapidly, such as "1,2 dimethylcyclooctene plus ozone followed by hydrolysis yields decane-2,9-dione." Students listen *without writing anything on their papers* this time around. Then, he repeats the reaction slowly, and students draw the appropriate compounds and reaction. Finally, he reads the reaction aloud a third time to give students a chance to make minor corrections.

"I'm trying to teach them to think in the language of organic chemistry," says Finley. By asking his students to "skip" the step of writing the words that represent formulae and/or reactions, Finley wants them to gain a more immediate relationship to that "language," a little like the leap made by the foreign-language student who suddenly finds his/herself "thinking" in the new language, not translating. Students apparently enjoy the *dictées*, which makes the learning process that much easier.

Before taking his course, many of Finley's students expect organic chemistry to be the "weeder" course for potential medical school applicants, requiring an almost impossible amount of memorization. "This is absolute nonsense," says Finley. "I don't teach my course that way, and I myself am living proof that you don't have to have a superb memory to understand organic chemistry. But I have to spend much of my time convincing my students otherwise."

GROUP PROBLEM SOLVING, MIXED-MODE ASSESSMENT

Bill Flurkey, Department of Chemistry, Indiana State University, Terre Haute, IN 47809; TEL: (812) 237-2245; FAX: (812) 237-2232; E-MAIL: chflurke@scifac.indstate.edu.

Courses Taught

- General Chemistry (for nursing students)
- Biochemistry
- Biochemististry for Dieticians

Description of Examination Innovations

Bill Flurkey offers his students a mix of exercises to test a wide variety of skills, knowledge, and abilities to apply knowledge in new settings. Among these are take-home exams, open-book exams, and problem-solving questions involving group discussion. One day each week, groups of four to five students (formed at the beginning of the semester) spend 15 minutes completing three or four review questions. One member of each group presents on the chalkboard the group's solutions to all the questions. After all group solutions have been provided, the class and Flurkey decide which group(s) provided correct answers for each question. Discussion usually follows to explain and to correct incorrect solutions. This year, he is trying smaller-size groups, and in the future, Flurkey may make these problem-solving sessions count toward a specified number of "groupwork" points that can be earned during a semester.

In the area of exam content, Flurkey requires students to describe the structure–function relationship of molecular models that he brings into class. A typical exam will have students draw structures, answer multiple-choice questions, and write answers to short essay–type discussion questions.

Flurkey says students like his varied approach to testing because those with different testing preferences get an equal chance to display their knowledge. Based on a comparison of past student evaluations with current student evaluations of the class and instructor, Flurkey believes attitudes toward biochemistry have improved.

TAKE-HOME ESSAY, ORAL PRESENTATION

Susan H. Ford, Department of Chemistry and Physics, Chicago State University, 9501 S. King Drive, Chicago, IL 60628-1598; TEL: (312) 995-2171; FAX: (312) 995-3809; E-MAIL: bij1shf@uxa.ecn.bgu.edu.

Courses Taught

* Biochemistry II—Chemistry 313

Description of Examination Innovation

Susan Ford uses two innovative exam practices in her Biochemistry II course. The first is a take-home essay, which counts for 20 percent of the first of two exams (the other 80 percent is graded on in-class essay writing). Essays from the take-home portion are carefully graded on content, style, and English usage, and then returned to students. Students may resubmit their essays after rewriting for a higher grade. Her rationale is this: "Since all examinations in this course are heavily essay type, students need extra practice early in the course in writing on complex scientific subjects." She has been using the regraded take-home exam for 2 years.

One disadvantage to the take-home exam/rewrite is the extra grading time required and the need to think up new and challenging topics for the assignment. The topics are suffiently broad and unique to deter copying from another source. Ford does encourage students to discuss possible responses but warns them that she personally reads all the papers, and that answers which are too similar will be suspect. With 10–15 students enrolled in the class, she has not, as yet, had a problem with cheating.

Ford's second innovation is a required oral presentation that counts as 25 percent of the final exam grade. Students choose a current biochemical topic from articles and papers the instructor supplies from recent journals and from the popular press. Students prepare a 25-minute lecture/discussion in which they are expected to "teach" the class. They can use overhead transparencies, slides, and handouts, which the instructor supplies, if requested. Both their classmates and the instructor award grades to each presenter. On the written final exam, the content of the student presentations is included, which forces students to concentrate when their peers are presenting.

Students, says Ford, appreciate the opportunity to experience the "other side" of college teaching, which for them is often an "eye-opener." They learn that teaching requires more organizational skills and thought than they had previously assumed. Since the biochemistry course is for seniors, students generally have sufficient background and maturity to perform well.

Ford says both innovations would be difficult to do in larger classes because of the extra time required for grading or class presentations. Still, at least two of her colleagues have incorporated these innovations into their courses.

FINAL ESSAY BASED ON ORGANIC COMPOUND

Dolores Gracian, Department of Chemistry, Bronx Community College, University Avenue and West 181st Street, Bronx, New York 10453; TEL: (718) 289-5556; FAX: (718) 289-6075.

Courses Taught

- General Chemistry
- Organic Chemistry

Description of Examination Innovation

In advance of her organic chemistry final exam, Dolores Gracian asks students to choose an organic compound with more than one functional group (subject to the instructor's approval) and to study it in depth for the test.

On the day of the examination, students must be prepared to write a short essay on their chosen compound. The essay should include answers to the following questions: (1) Why are you interested in this compound? (2) What is its present use or possible future applications? (3) How is it prepared or extracted from natural resources? and (4) How could it be modified using reactions you have studied?

Students are permitted to bring into the exam room one index card describing the structure of the compound, its physical constants, a short outline, dates, and other references. This way, Gracian feels her students focus more on understanding their organic compound than on memorizing the features of a list of compounds. Because her students have already passed a first semester of organic chemistry and taken two of her second-semester lecture exams, Gracian feels this exam lets her assess what she feels is most important for her students: the ability to know how to get needed information and to apply what they know to a given situation.

She has been giving this exam for 4 years. Students say they enjoy finding out this in-depth information, especially those planning to go into medical fields.

ORAL EXAMINATIONS, STUDENT-GENERATED QUESTIONS

Iclal Hartman, Department of Chemistry, Simmons College, The Fenway, Boston, MA 02115; TEL: (617) 521-2730; FAX: (617) 521-3199; E-MAIL: ihartman@vmsvax.simmons.edu.

Courses Taught

- Biochemistry
- Environmental Chemistry
- Chemistry of Drugs and Drug Action

Description of Examination Innovation

In her small upper-division chemistry courses at Simmons College, Iclal Hartman has established a tradition of giving oral examinations. Recently, Hartman decided to retain the oral exam format but to make it more interactive and less question–answer oriented. For her larger classes, she invites students to participate in the design of their written examinations.

Oral examinations are scheduled in 30-minute segments for "hour" exams and in 1-hour blocks for finals. It is possible, says Hartman, to extend the oral exam format to larger classes of up to 24 students by pairing them. Pairing works best when the two students are of about equal academic standing in the class. When carefully planned and executed, the oral exam format can be both interactive and cooperative, a forum for testing as well as for an exchange of ideas between students and between student and teacher.

For topics that require extensive calculations, graphing, and so forth, the instructor usually gives supplemental written exams. However, she also finds it possible to include such topics on the oral by having students set up a logical equation or formula on the chalkboard without taking the time to do the calculations to arrive at a numerical answer. The approach enables students to appreciate the primary importance of the correct approach and logical set-up for the solution of a problem over the mechanics of calculation.

To further facilitate an interactive discourse, the instructor supplies students during the oral exam with any factual information they might need. This makes it clear that she is not testing memory. Each student is free to bring his/her laboratory notebook, lecture notes, and textbook to the oral exam, and is provided with a handbook and other source materials to consult as needed. Although the instructor is at pains not to minimize the importance of a broad foundation of basic (factual) knowledge, she conveys to students her firm belief that such mastery will (and should) come in time.

Her aim, as she puts it, is the cultivation of mental, not memory, power. "Whatever be the detail with which you cram your student," she believes, quoting Alfred North Whitehead, "the chance of his meeting in after-life exactly that detail is almost infinitesimal....[Better] a training [that] yields a comprehension of a few general principles with a thorough grounding in the way they apply to a variety of concrete details."[4]

In oral examinations, Hartman tries to follow the Socratic mode of inquiry. Even when identical questions are given, the discourse follows a different path with each student. Once students get over their initial nervousness at the less familiar format of the oral, it is easy for the instructor to assess whether the material has been mastered and can be applied in fresh combinations. Hartman must also balance their answers against the amount of assistance they needed, and the detours they took, although figuring out the answer—however arrived at—is always a plus. If a student doesn't know the answer, she does not let the student get mired in confusion but quickly moves on to a different topic.

Students have not challenged oral exam grades; in fact, many say that they assumed they had not done as well as their grade suggests. The instructor believes that "the oral examination is an excellent and immediate teaching tool." Its only disadvantage is scheduling and some students' heightened anxiety over an oral interaction with their instructor.

A second innovation by Hartman has been to involve students in the cooperative process of constructing a written examination by inviting them to submit questions. Three purposes are addressed: (1) to help students in their study process to organize and correlate the material, recognize important concepts and principles, and understand their interrelationships and applications; (2) to help students learn to formulate questions with clarity and precision within a defined context; and (3)

---

[4] Quoting Alfred North Whitehead in Iclal S. Hartman, "Interactive and Cooperative Methods as an Extension to Examinations," *Journal of College Science Teaching* (May 1995): 402.

to aid students when they are asked to supply the complete answers to their questions. Hartman takes the student-generated questions, which tend to be isolated and somewhat specific, and tries to organize them into segments of broader questions. She retains students' wording, changing only for clarity or uniformity of form within the larger question. Good-quality questions contribute to the 5 percent of the course grade that is based on class participation and improvement over the term. Poor questions have no impact on grades.

The instructor finds that students are "surprised," then "pleased" and "challenged," by the role reversal involved in writing their own exams. Those who do the assignment (about half of the class) take it seriously. The cooperative experience gives students more confidence in asking questions they might earlier have considered "stupid," or in exploring a "wild" idea. Overall, the innovation enables students to recognize their role as active partners in the teaching–learning process and to appreciate the care that goes into constructing a comprehensive, challenging, and unambiguous in-class examination.

MULTIFACETED GRADING OF LAB WORK

Marjorie Kandel, Department of Chemistry, State University of New York–Stony Brook, Stony Brook, NY 11794-3400; TEL: (516) 632-7945; FAX: (516) 632-7960; E-MAIL: mkandel@ccmail.sunysb.edu.

Courses Taught

- Organic Chemistry Laboratory
- Intensive Organic Chemistry Laboratory
- Elementary Chemistry Laboratory

Description of Examination Innovation

A grading system that evaluates skills directly is used by Marjorie Kandel and colleagues in their large organic laboratory course.[5] To measure their students' ability to solve lab problems and perform manipulations, they grade products. To measure attention to the progress of the experiment and faithfulness in keeping the record, they grade notebooks. To measure the knowledge acquired and ability to apply it to lab problems, they grade traditional exams and a final

---

[5] The description of the innovation is taken (with the author's permission) from two articles both by Marjorie Kandel: "Grading Laboratory Notebooks in a Large Organic Chemistry Course," *Journal of Chemical Education* 63(1986): 706; "Grading a Large Organic Laboratory Course," *Journal of Chemical Education* 65(1988): 782.

report. And to account for the immeasurables that nevertheless may be apparent to the instructor, they have a small technique grade.

The product grade (30 percent of the total) has both a qualitative and quantitative component, the first being emphasized. Student-supplied data are not used, but, rather, the instructor evaluates the sample's appearance, melting point, if solid, and gas chromatogram if liquid, and weight or volume. Because it is apparent that doctoring of samples can be detected, there is little tampering. Students rarely complain about a product grade, the instructors report, because it is perceived as objective.

Notebooks are given the same weight as products in the grade scheme. To ensure that students write as they work in lab, the instructors collect carbon copies of each day's pages at the end of the period. During the semester, several short, open-notebook exams are given with questions on procedure, data, observations, and those conclusions that must be made for the experiment to be continued. Grading notebooks by this type of exam is far less time-consuming than by traditional methods, and, the instructors report, grades are less frequently challenged by students.

Forty percent of a student's total grade is determined by theory–practice exams and a final report. The open-book exams call on students to interpret typical data, some of which are their own. The interpretive role of traditional lab reports is thus assumed by these exams. However, to encourage writing, the final report, an essay on the qualitative analysis of an unknown compound, is retained. Students are given a brief outline for guidance and a page limit. Part of the grade is based on the quality of presentation, including grammar, spelling, and editing. Grading of the final report is time-consuming, requires mature judgment, and is the responsibility of the course's most skilled TAs.

MODIFIED MASTERY LEARNING SYSTEM

Marlene Katz, Chemistry Department, St. Louis College of Pharmacy, 4588 Parkview Place, St. Louis, MO 63110; TEL: (314) 367-8700 Ext. 241; FAX: (314) 367-2784; E-MAIL: katzm@medicine.wustl.edu.

Courses Taught

- Organic Chemistry
- Biochemistry

Description of Examination Innovation

Marlene Katz has found a way to incorporate mastery learning into her organic chemistry classes while still giving students the choice—if they prefer—of being graded conventionally. Three grading options are given to students for the "specific skills" quizzes 8 to 10 times a year: mastery, modified mastery, and conventional. Mastery quiz grading allows the student a total of five attempts to pass the quiz, with a passing grade given for work without any significant errors (nearly perfect). Modified mastery grading allows the student to repeat a quiz once and have the two scores averaged. In the conventional system, quizzes are graded by scoring the percent correct. This option does not allow retesting.

Katz administers and grades the first quiz of the semester three ways, once according to each system. Then she asks students to decide which grading method they will abide by during the term. The skills tested are those she considers central to the course, and she tries to separate the material out so that each quiz contains hard, moderately difficult, and easy questions. Quizzes are short, with approximately four questions, some requiring student drawings.

To prepare students for mastery learning, Katz distributes a skills list for each quiz in a packet of supplemental materials distributed at the beginning of the semester. Skills include the following: "Know the reagents required for functional group interchanges of alcohols," and "Be able to draw the cation intermediate with its resonance forms for electrophilic aromatic substitution." Also in the packet is a sample quiz that Katz will check at any time, once completed. About 25 percent of the students avail themselves of this option, and others in study groups also review the corrected sample quizzes. She also reviews completed sample quizzes with common student errors.

This is a labor-intensive testing system, says Katz, especially in the preliminary stages. "The major work involved is deciding which skills are central to the course and which skills should be tested on each quiz," she says. Katz writes the masters for all quizzes and retests before the semester begins. Students look at their graded quizzes during one of the help sessions or during her office hours. Because Katz does not release completed quizzes, they can be used again, but each year, she adds a few new retests to her sets.

Katz is willing to put in the extra effort required to prepare and grade these quizzes, because she sees positive results in achievement and retention. Her students respond positively to mastery learning and usually elect either mastery or modified mastery. In fact, this semester, *none of her organic chemistry students opted for the conventional system of grading.*

Katz has been using this method for 5 years at Huntingdon College, a small liberal arts school, and for 2 years at the St. Louis College of Pharmacy. She instituted the system at Huntingdon after a year when many bright students performed poorly on the cumulative final exam. She hoped to increase retention of course material by using a system that promoted mastery rather than minimal learning to get a desired grade.

Student reaction at both schools has been similar. Some students are intimidated, believing themselves unable to achieve near-perfection. Others are relieved of test anxiety by the opportunity to retest. An end-of-semester survey showed that the majority of St. Louis students were happy with the quiz-grading plan they chose.

Recently, because none of Katz's students were choosing the conventional grading system, she opted to offer only two quiz-grading options. One assigns quiz grades on a pass–fail basis, with a retesting option and the number of "passes" determining the overall quiz grade. The other gives quiz grades on a percent-correct basis, with the retesting option and the overall quiz grade determined by the average percent calculated from averages of retests for each quiz.

GROUP PROJECTS USING COMPUTER SPREADSHEET

R. Gerald Keil, Associate Dean, University of Dayton, 300 College Park Drive, Dayton, OH 45469-0800; TEL: (513) 229-3135; FAX: (513) 229-2615; E-MAIL: KEIL@mccoy.as.udayton.edu.

Courses Taught

- Quantitative Chemical Analysis
- Physical Chemistry
- Electrochemical Methods

Description of Examination Innovation

"Suppose you wanted to be an excellent downhill skier?" Gerald Keil asks his quantitative chemical analysis students. "Would you be better trained by viewing the videotape of a professional skier taking the slopes over and over? Or would you watch the videotape once and then try some skiing yourself?"

Keil and his students know that people learn better by doing rather than watching, which is why he has his students "practice" doing analysis with computer-generated spreadsheet problems. Indeed, 70 to 75 percent of the graded work in Keil's class is based on group projects employing spreadsheets. Over the last 5 years, spreadsheet programs such as Quattro Pro and Lotus 1–2–3, on a 386 IBM-compatible computer, have become increasingly popular in chemistry classrooms. Keil has made his spreadsheet assignments an integral part of his evolving pedagogical style.

At least half of his lecture time is now allocated to group projects using spreadsheets. Halfway through the period—sometimes earlier—the class breaks up into its working groups of no more than four students. While students are working in class, Keil walks around the room engaging them. In course evaluations, students say they appreciate the opportunity to get so much feedback. Keil understands this.

"In a traditional lecture class, the only opportunity students get for feedback are the grades on three or four exams," he says. "This way, if the students are working, they get feedback in almost every class." With the changes in Keil's pedagogy, students are held responsible for understanding the chemistry in the textbook. Their success with the spreadsheet projects tells them (and him) how well they're doing.

"The beauty of the spreadsheet is that it allows a myriad of calculations to be done very easily," says Keil. "The computer does all the work, so what's left up to the student is figuring out the process: Which equations should be used, and why? This is the most important learning objective and course outcome. And since they know, from the laboratory, what the titration curve looks like, if part of the curve 'doesn't come out right,' their job is to go back and figure out what's was wrong with their reasoning." Since students don't have to worry about calculation errors with the computer, they can think more deeply about the process.

Most of Keil's spreadsheet problems replicate work done in the lab and, therefore, reinforce that work. For example, a typical spreadsheet problem asks students to consider marble statues and the effects of "weathering" from pure and acidic rain. Students must calculate the time required to remove 10 percent of the statue mass as a function of acidity of the rainwater. They also calculate the pH to which deionized water equilibrates, using text materials and spreadsheets. Finally, they must find the time required to damage the marble at different acidities. Extensive graphing and calculation, as well as spreadsheets, are used to explore the weathering process.

Keil says it's rare that anyone earns less than 90 percent on any one spreadsheet assignment. This counts for 75 percent of the course grade. But the other 25 percent of the course grade is derived from "very easy" exams based on the spreadsheet problems. "If students have done the spreadsheet problems, they'll do well," says Keil.

When asked why he uses cooperative learning with the assignments, Keil says he felt the need to get more of his students engaged in learning the material, and lectures were failing him in this regard. In science education journals, he kept reading articles with variations on the title "What do you do when you stop lecturing?" Eventually, Keil decided to switch to cooperative learning.

"As I see it, most students today watch a lot of TV," Keil says. "I feel my students, like most, expect everything they do, including learning, to flow easily. They're just not used to getting involved in difficult projects, and the only way I know how to do get them actively involved in my class is to have them working in teams."

Twenty percent of his students, however, typically prefer to "go it alone" on the projects. The final exam is always an individual, in-class exam. He tells them: "You don't have to know every aspect of the course, but a demonstration of critical-thinking skills consistent with the subject is expected. You are here to ask a lot of questions and learn."

Initially, students are "taken aback" when told how extensively they'll be using the spreadsheet for the course. But by the end of the semester, they're grateful. They say they're learning the material better through the group work. Keil's colleagues, who meet Keil's former students in higher-level courses, are pleased, too, that his students have learned to use spreadsheets. In fact, they've told Keil that his former students turn to the spreadsheet on their own in physical chemistry and other course work.

Most important, though, Keil feels he's reaching more of his students today than he ever was before. "With the spreadsheets, we're doing more difficult problems," he says. His students are asking more in-depth questions in class. "It's not a 'cake' course," says Kiel, and his students get a range of final grades, but he thinks the scope and depth of the course have increased.

ORIGINAL TEST-ITEM BANK, COMPUTERIZED GRADING

Joseph Lechner, Department of Chemistry, Mount Vernon Nazarene College, 800 Martinsburg Road, Mount Vernon, OH 43050-9500; TEL: (614) 397-6862 Ext. 3211; FAX: (614) 397-2769.

Courses Taught

- General Chemistry
- Organic Chemistry
- Biochemistry

Description of Examination Innovation

Joseph Lechner is a staunch defender of some traditional testing strategies and is eager to demonstrate that multiple-choice tests, test banks, and computer-grading software programs can provide test diagnostics for both students and their instructors.

Six years ago, Lechner realized he was rewriting the same exams year after year, so he started putting his multiple-choice questions on 3" × 5" cards and selecting cards from the bank to create new exams. He also began including statistical data, such as percent of correct responses and point-biserial R, on the backs of the cards to increase the reliability of the exams he created. The cards are filed in metal cabinets, with a different bin for each day's lecture topic. The topic is coded on the back of each card.

At about the same time, he began reorganizing an outdated software grading program so that many more helpful test diagnostics were available to the instructor, such as a feature that correlates lecture topic and day with test items and results. His policy is to include at least five test questions from each day's material. After the exam has been administered and the answer cards have been scanned, Lechner types in the title of each day's lecture straight from the syllabus and identifies all the questions in the exam that pertain to that day's lecture. The program gives him a printout by day, including every test item based on that day's lecture, and how the class did on that test item and on all items pertaining to that day. This helps Lechner determine what his students are not understanding.

Each student also receives a list of topics covered on the exam and an indication of how well he/she did on each day's topic—valuable information, as Lechner's students are responsible for mastering material they don't understand on an exam. It is the instructor's standing procedure to include on subsequent exams similar, though not identical, review questions about the same material from a previous test. He uses test-item analysis to learn which topics need to be retested, and then 10 percent of the succeeding exam consists of review items.

The software also allows Lechner to classify questions according to Bloom's taxonomy of instructional objectives (his questions are designed to employ all six levels) and to detect incorrect answer keys. The program scans the answer cards on a hard-disk file for all the other programs to look at. The program also enables him to check for cheating.

Although creating a test-item bank takes time in the beginning, Lechner says the system now saves him a great deal of time. There is a quantifiable advantage for students, too, in that by having a ready supply of trusted exam questions, Lechner's exams tend to be longer (60–70 items instead of 40–50; longer exams increase students' scores) and more reliable. (Kuder–Richardson coefficients, which measure overall exam reliability, are at the .91 or .92 levels.)

In his more advanced biochemistry classes, Lechner supplements multiple-choice with essay questions. But at the level of organic chemistry, Lechner insists he can "do it all" with multiple-choice questions, including speculation based on experimental results. Lechner feels that faculty should not be resistant to multiple-choice questions, particularly when their students face GRE and MCAT exams that use multiple-choice questions to test higher-level thinking.

"I feel strongly that content and critical thinking can be taught through carefully constructed multiple-choice items," says Lechner. "Canned" test banks from textbook publishers meet with his disfavor, however. The thought that goes into creating one's own test is beneficial for the instructor.

Lechner recently presented his methods to the faculty at Mount Vernon so that other professors could improve their exams and minimize the time they spent working on them.

GROUP DISCUSSION BEFORE INDIVIDUAL EXAMS

G. Marc Loudon, School of Pharmacy and Pharmacal Sciences, Purdue University, W. Lafayette, IN 47907-1339; TEL: (317) 494-1462; FAX: (317) 494-7880; E-MAIL: loudonm@omni.cc.purdue.edu.

Courses Taught

- Organic Chemistry (for pharmacy majors)

Description of Examination Innovation

To combine group work and individual grading, Marc Loudon and then-graduate student Richard Bauer have designed an interesting format for exams. Students are given a copy of the exam to work on in their study groups. Forty-five minutes are allotted for discussion of strategies for solving the problems and for coming up with specific answers. Then, their written materials are collected, and each individual student is given a clean copy of the exam to work on alone.

Because of the new exam "ecology," the instructors find that they can ask questions at a more sophisticated level than what normally would appear on sophomore-level organic chemistry exams. This means, in turn, that they can present material in class at a higher level. There are no multiple-choice, true–false, or other retention-type questions. Rather, typical exam items are open-ended, and the problems the instructors select require students to explain their reasoning as they solve them.

In grading exams, Loudon uses a "resurrection" grading system (see Part 1, p.100) only for calculating the final grade for the course: Poor midterm exam scores can be replaced by a good final exam grade.

Initially, the new examination method was tried out experimentally with a randomly selected group of 40 students from the large sophomore-level course. It was accompanied by a new nonlecture format of presentation as well. Considerably less material was covered owing to group discussion, which was employed during class to answer questions the instructor posed. Because of this, students were held responsible for material that was assigned but not specifically covered.

Some of Loudon's colleagues were "aghast," he reports, at the idea that students could openly discuss questions on which they would be graded individually. Loudon and Bauer countered this kind of criticism by reminding colleagues that this is the way they solve problems. They argued that students need to learn that science is a social endeavor requiring collaboration, and that real-world chemical problems are never multiple choice. In Loudon's view, "The typical examination format is artificial, lifeless, and virtually irrelevant." Long ago, he grew weary of having students tell him (as if the two were incompatible), "I'm not going into research, because I want to work with people."

Proof that his system works is that when the experimental section was integrated into a traditional section for the second-semester course, as a class, they performed 25 points (out of 400) better than the other students. Of the group, 32 percent earned A grades, compared to 16 percent of the traditionally taught and examined students; none failed, and there was only one D (in comparison to the traditional section, where there were a significant number of D and F grades).

The major change, in addition to enhanced performance, was one of attitude. There was less anxiety, and the instructor had more fun getting to know every student personally, although he found he did not have to provide much personal in-office assistance to students in the special section.

Two vignettes deserve to be related. One is that two students (both female) asked to be switched out of the special section because they didn't want their grades depending on other people. This did not, however, lead to the "mass exodus" the instructors had feared. The other is that one of the special-section students became an informal tutor for a student in a traditional section. The latter student wanted to gain some of the former student's deeper understanding.

COOPERATIVE TAKE-HOME EXAM

Theodore L. Miller, formerly of Ohio Wesleyan University, Battelle Medical Research and Evaluation Facility, 505 King Avenue, Columbus, OH 43201-2693; TEL: (614) 424-4790; FAX: (614) 424-3317; E-MAIL: millertl@battelle.org.

Courses Taught

- General Chemistry
- Analytical Chemistry
- Instrumental Analysis
- Science courses for nonmajors

Description of Examination Innovation

In the 1992–93 academic year, when he was teaching analytical chemistry at Ohio Wesleyan University, Ted Miller made an interesting discovery about testing. He was in the process of using cooperative learning and special techniques, and his fall semester began as all of his classes do: a lot of time was spent making students comfortable with their active learning groups. And then, 3 weeks into the term, Miller gave students a traditional take-home exam to complete individually, explaining that no student was to speak with another about the exam.

"The atmosphere in class totally changed," says Miller. "The sudden switch from cooperative learning to individual assessment troubled the students." After discussing the experience, he and his class agreed to try a cooperative take-home midterm (using the same format as the individual take-home exam).

Miller says he learned that "cooperative assessment, like cooperative learning, enhances the learning of chemistry." Since this experience, he uses cooperative assessment for all his analytical chemistry and instrumental analysis midterms. The results, he says, are more than promising: He administered the standardized exam prepared by the American Chemical Society as a final in both courses, and the average percentile ranks were 81 and 73, respectively—the average percentile rank for 469 students in traditional lectures courses between 1974 and 1991 was 55. Miller also uses quizzes on basic concepts, papers, and standardized final exams to determine individual accountability.

"For me," says Miller, "these results clearly illustrate the advantage of cooperative learning groups with cooperative assessment." From his methods, a student said she had learned how to evaluate "my strengths as an overall student."

Miller wasn't always the avid practitioner of cooperative learning that he is today. His teaching style changed slowly. Now his chemistry classes revolve around problem solving, decision making, and social values in the context of chemistry, and his students' work is focused on active learning. This takes place in semester-long, small-group projects that include a summary paper, field trip, oral report, and unit quiz. Students "contract" for a passing grade by agreeing to do different amounts of work at least 90 percent successfully.

For Miller, the biggest obstacle in using cooperative assessment methods was his concern for fairness to individual students. Indeed, most of his colleagues remain skeptical of such practices for that reason.

VARIANT OF SELF-PACED MASTERY LEARNING WITH COMPUTERIZED QUIZZES

Peter J. Moehs, Department of Chemistry, Saginaw Valley State University, 7400 Bay Road, University Center, MI 48710; TEL: (517) 790-4349; E-MAIL: pjm@tardis.svsu.edu.

Courses Taught

- Qualitative Analysis
- Instrumental Analysis

Description of Examination Innovation

Since students at Saginaw Valley State University, primarily a commuter school, have less time for studying, Peter Moehs has experimented with semisupervised computer quizzes (with randomly selected questions) that his sophomores and juniors are allowed to do at their own pace. Each quiz is designed to cover a selected section of the course. Only the best of three tries on each quiz is recorded as the student's final quiz grade.

The beginning quizzes review concepts students should have mastered in lower-level courses. Subsequent quizzes progress to subject matter covered primarily in the qualitative and instrumental analysis courses. For example:

A 40 percent solution of KOH has a density of 1.42 g/mL. What is the molarity of this solution? (Given: F.W. KOH = 56.1 grams/mole.)

OR

Fiber optics is finding increased applications in instruments as signal transmitters and miniature sensors. If n1, n2, and n3 are refractive indices of the fiber, coating, and analyte, respectively, which of the conditions produces the greatest aperture angle?

If a student does not obtain a successful grade on any one of the quizzes, he/she may obtain one-on-one instructor assistance with a quiz set. Inexpensive Apple IIe computers are used to run this Keller-type student learning module. Grading is done by computer as students take the quiz. Their scores and tabulated results on each quiz are recorded on the data disk given to them for each attempt.

When Moehs first started giving computer-graded quizzes in 1987, there was a fair amount of apprehension due in part to students' lack of familiarity with computers. But in the last 5 years, students seem to welcome the technique, perhaps in part because most of Saginaw Valley's chemistry majors have part-time cooperatives at corporations such as Dow Chemical Company and Dow Corning Corporation (both situated close to the institution). Since 1990, the instructor has heard few negative responses related to students cheating. Other faculty find the extra time forbidding, though they welcome Moeh's efforts.

Moehs reports his class averages rose 10–15 percentile points with the microcomputer quizzes, as measured by the standardized American Chemical Society examinations.

FINAL EXAM SUPPLEMENT

Dave Reingold, Department of Chemistry, Juniata College, Huntington, PA 16652; TEL: (814) 641-3565; FAX: (814) 641-3685; E-MAIL: reingold@juncol.juniata.edu.

Classes Taught

- Organic Chemistry

Description of Examination Innovation

Dave Reingold has been doing some experimenting with examination formats in his organic chemistry classes. Noteworthy is his "final exam supplement," which he appends to the students' standardized multiple-choice exam. (The final is the only exam in which Reingold uses a multiple-choice format.) It is both a statement of his philosophy in regard to the limits of examinations and an opportunity for his students to "show what they know." His supplement is reproduced below in full:

> For each exam, you study a great deal of material, only some of which appears on the exam. It is anybody's guess whether the things you know best will be asked. Sometimes I ask what you know, and you do well, and sometimes I ask what you don't know, and you do poorly. Often I ask things which you think you know, but the question is phrased in such a way that you get flustered and don't do well. Here is your chance to show me that you really learned something about organic chemistry this year. You have just finished answering a bunch of short-answer questions about organic. Now sit back, relax, and think of some aspect of organic that you think you know well, and convince me of it.
>
> Starting on the next page, write a several-page essay about something. Your discussion should include at least one place where you draw a mechanism, and at least some discussion of evidence (i.e., How do we know what you are saying is true). This should not be a recap of some question you have seen earlier, either on an exam (this year or earlier) or in the book. Please write clearly.

The supplemental question is worth 40 percent of the final exam grade. Most students like the question, but Reingold uses it sparingly. He says if students knew it was coming, they might forego studying other aspects of the material than what is covered in the standardized section of the final. They might also prepare an essay in advance, which isn't the instructor's intention. As an occasional technique, the question can be used to advantage in any course, says Reingold. It gives students an opportunity to be tested on what they know, something students complain doesn't happen on standardized exams.

QUALITATIVE ASSESSMENT OF STUDENTS' PROGRESS

Giacinto Scoles, Department of Chemistry, Princeton University, Princeton, NJ 08544; TEL: (609) 258-5570; FAX: (609) 258-6665; E-MAIL: gscoles@pucc.

Courses Taught

- Thermodynamics
- Chemical Dynamics
- Physical Chemistry

- Solid State Physics
- Laboratory Courses
- Material Science

Description of Examination Innovation

Giacinto Scoles's innovation in grading practice is one of the most radical in this collection. He does not give letter or numerical grades in chemistry classes with 20 students or fewer until the final course grade is assigned. Somewhat of an iconoclast, Scoles feels most college professors rely too heavily on examination scores.

"We all know that examinations are a kind of measurement, and as such, they are subject to errors," he says, "yet many college educators feel they ought to ignore the information they've culled about student abilities and progress throughout the semester and use only the examination grades."

Twenty-five years ago, Scoles looked at the realities of college science exams and decided to do something. If written examinations were only one form of measurement of student progress, then he felt, as a scientist, he needed more "input data." He made a commitment to gaining as much information about each student as he could during a term. From that point on, he abandoned traditional grades based solely on "dry outcomes."

At the beginning of the term, Scoles carefully explains his grading system to his students, including his specific expectations for an A, B, C, or lower. He tells his students that the best among them will be those who: (1) organize a poorly defined problem into a well-posed question; (2) obtain rapidly the information necessary to solve a problem when necessary; and (3) generate new ideas (where new is not new *per se*, in an absolute context, but may only mean new in the student's context). As in most college science courses, weekly problem sets are assigned, which Scoles corrects and returns without grades. Students are supposed to solve the problems on their own. Asking peers or the instructor for help is allowed; having someone else solve the problem is not. A conventional honor code is enforced.

Scoles also collaborates with individual students at least once during the semester, meeting with them over 2–3 weeks to work on a project and on more difficult and/or time-consuming problems.

Scoles feels in this way he learns more about the nature of student errors (calculational or conceptual), and about their level of motivation and ability in the class. By getting to know each student's strengths and weaknesses, he feels his final judgment, which is in part the basis of the student's final grade, will be more accurate than a grade based only on a couple of written examinations.

But how are grades assigned? As is made clear at the beginning of the term, Scoles meets individually with each student at the end of the semester to see if a final examination is needed. During this meeting, he and the student each "grade" the student's performance during the semester by writing down either a letter or a numerical grade. There is no need for an exam if there is 10 percent numerical or letter agreement within what Scoles calls "one notch"—A– and B+

are considered in agreement and the final grade is that given by professor, whereas A– and B are not in agreement.

When disagreement results, no matter who gives the higher grade, some type of final examination is set up and administered. Sometimes, a longer problem set is given; other times, an oral exam might be given—say, if the student claims he/she has strong conceptual understanding but is weak in calculational ability. If necessary, Scoles will have a colleague mediate by participating in an oral exam or devising an assignment that will resolve the discrepancy. Scoles carefully avoids negotiating with students, which he feels would jeopardize the integrity of the system.

"What's the point of using an examination measurement to assign a course grade," asks Scoles, "if the student and I agree on a grade?" The difference, in Scoles's system, is that assessment is based on qualitative information about students' progress. "Involving students in the decision is also a sure way to get them to buy into the system," he adds.

The method appears to be working. Students are "very diffident" at first, but on average, no more than 10–15 percent of the students request a final examination. Most are "surprised and pleased" with the results. (Half the time, the students give themselves lower grades than their professor.) They are motivated, says Scoles, to be objective in their self-assessments, because they would like to avoid taking a final exam.

Scoles has been using this method for over 25 years in two classes per year. "For the system to work, the instructor has to get to know each student quite well before the course is over," he says. This requires giving attention to students and probably requires a small class size. In a sense, the whole course becomes an "examination."

According to Scoles, other professors feel this method is extremely interesting, but so far, no one he knows has adopted it because it requires more time than standard methods and, possibly, because in a few extreme cases his system requires breaking some bad news to students in one-on-one meetings. Writing a numerical score at the end of a written exam does not require this type of direct confrontaton.

DAILY GROUP QUIZZES, DEMONSTRATION-BASED QUIZZES

Robert G. Silberman, Chemistry Department, SUNY at Cortland, Box 2000, Cortland, NY 13045; TEL: (607) 753-2912; FAX: (607) 753-2927; E-MAIL: silberman@snycorva.cortland.edu.

Courses Taught

- Introductory Chemistry

- Organic Chemistry
- Modern Chemistry in the High School Curriculum (for high school teachers)

Description of Examination Innovation

For 2 years, Robert Silberman has been giving short, cooperative quizzes at the end of each class period that he thinks increase student learning in his organic chemistry classes.

At the end of each lecture, he writes a one- or two-part conceptual problem on the board, based on the material he just covered, often making it up on the spot. Usually, the problem requires students to write a few sentences or make a drawing explaining a concept, outline a synthetic scheme, or propose a mechanism for a reaction.

For example, one quiz asked students the following question: "Explain why halogens are ortho-paradirecting groups, but ring deactivators. If possible, use drawings in your explanation."

Students work on the quizzes for 5–10 minutes in groups of three, while Silberman circulates through the room observing the class. "As I look around the room and hear all the arguing, I know my students are working hard," he says. "I also have a feeling they're paying better attention during lecture, because they anticipate a quiz. When we go over the quiz, it serves to review and reinforce what they have just been learning."

The quizzes account for only 10 percent of the course grade, so they are not punitive. Silberman is not trying to frighten students into listening better. He is rewarding those who do as he guides them through the learning process.

Although quizzes almost every class period generate a lot of paperwork, Silberman says that in classes of 60–80 students using only one or two questions on a quiz, it is not prohibitive. In surveys, students say they like the quizzes—in part, no doubt, because they tend to boost the grades of most students. That is not a problem for Silberman. "They are learning more. Nor is it grade inflation," he says. The quizzes also serve as a review, emphasizing an important concept in the day's lecture.

Silberman uses quizzes in other interesting ways. For example, he gives quizzes or portions of exams based on a particular demonstration. Students are shown a demonstration in class twice without being given any chemical explanation. Four to 5 days later, they are to hand in a short report explaining the chemistry or chemical phenomena involved in the demonstration. Students may do research in the library, but they are on their honor not to ask anyone for help. The write-up counts as a quiz or as 10 percent of the next hour exam. "I was surprised the first time I did this. Only one person in the class got it right," says Silberman. This sort of quiz helps teach students how to make observations, draw conclusions from those observations, and learn to use scientific resources.

QUALITATIVE QUESTIONS

Wade Sisk, Department of Chemistry, University of North Carolina–Charlotte, Charlotte, NC 28223; TEL: (704) 547-4433; FAX: (704) 547-3151; E-MAIL: wsisk@unccvm.uncc.edu.

Courses Taught

- Physical Chemistry (lecture and lab)
- Chemical Dynamics

Description of Examination Innovation

Wade Sisk aims for a quantitative/qualitative mix on his exams. Although physical chemistry involves quantitative problem solving, he usually asks students to answer qualitative questions (based on a quantitative knowledge base) for about 25 percent of the exam. On quantitative exam questions, he may ask for a quantitative answer on part A of the question and then ask for a justification as to why the problem could be solved without additional information. For example, part of a four-section question reads as follows:

> For the elementary reaction $OH + CH_3F$—products with rate constant $k1$: (a) For 2 points, write the differential rate expression for the depletion of OH in terms of $k1$ and reactants; (b) for 3 points, describe how to quantitatively determine $k1$: (i) what relative concentrations of OH and CH_3F; (ii) what (Y vs. X) quantities to plot; (iii) how to extract $k1$ from such a plot.

Students are asked to describe the quantitative determinations without doing them. Sisk likes these questions because they often demand more integration of knowledge and tend to be more comprehensive and less straightforward than purely quantitative questions.

Sisk started using these test questions 3 years ago because many of his chemistry majors were convinced chemistry was esoteric and had little application to the real world. Thus, some of the qualitative questions put the problem in context, so that it does not simply involve solving an abstract mathematical problem.

Since students want to know how to answer the qualititative questions, particularly how much to write, the instructor has learned to be quite specific in his directions. For example, to answer a question about the central theme in quantum mechanics as it relates to early experiments, Sisk will list for them some of the ideas he expects them to include, such as Planck's blackbody radiation formula, Einstein's photoelectric effect, and the Ryberg formula for H atom lines.

Over the two-semester course, as students become more accustomed to Sisk's style of questioning, they significantly improve at figuring out for themselves what they have to cover in answering qualitative questions.

ORAL PRESENTATIONS ON RESEARCH, LAB QUESTION, MAKEUP QUESTIONS

Bob Sweeney, Department of Chemistry, Fairmont State College, 1201 Locust Avenue, Fairmont, WV 26554; TEL: (304) 367-4498; FAX: (304) 366-4870.

Courses Taught

- Physical Chemistry
- Introduction to Chemistry
- Photochemistry

Description of Examination Innovation

In physical chemistry, Bob Sweeney uses those testing techniques that involve higher-order thinking skills. These and other important skills are demanded at every step of evaluation.

In one assignment that generates two student grades, Sweeney gives each class member a specific research article from a chemical journal such as *Environmental Science and Technology*. After students spend several weeks reading and analyzing the article, the next two class periods are devoted to 20-minute oral presentations in which the students explain the paper they were assigned. The other students ask questions and take notes, with the instructor occasionally asking for clarification for the benefit of all. After the presentations have been completed, Sweeney devises a quiz that includes one summary question from each presentation, some with data from a presentation reproduced on the quiz. A student's oral-presentation grade is determined by how well others in the class score on the question derived from the presentation. (The two lowest scores by others on the question are dropped.) The other grade they receive, the quiz grade, represents how well they absorbed the material they heard.

Sweeney says this technique "provides students with experience analyzing complex papers, giving oral presentations, and actively listening to one another." He has used it for 4 years in his physical chemistry classes. Although students at the time do not enjoy making the presentations, former students often comment on how useful the experience was for them. The instructor points out that chemists often have to give presentations, and that the classroom is a relatively safe place for students to gain the experience.

Also in Sweeney's testing repertoire are creative and challenging reorganizations of traditional laboratory work. For 4 years, Sweeney has made his students responsible for devising procedures to accomplish experimental goals with which he charges them, for example: "Determine the rate law for photochemical formation of the product of pyridine hydrolysis and determine the identity of the colored product." Half of the students' laboratory grade is based on the quality of their laboratory report, and the other half on the quality of the procedure they designed. Students are enthusiastic about this approach. "Designing our own labs makes us really think about the material instead of just repeating someone else's procedure," says one student. "I no longer have any doubts about my abilities in the lab," says another.

Sweeney also gives an exam question based on laboratory work, requiring students to evaluate the experimental procedures designed by previous students from the same class 2 years ago. He provides the class with the former students' detailed laboratory reports.

"The assignment is to critique the procedure," he says, "and find and correct major and minor errors in logic, procedure, and data analysis. This is quite a challenge for my students because they are so seldom called upon to use evaluative skills."

The students are given the same data as were collected by the students who wrote the report. Sweeney tells his students to analyze these data themselves and come up with results they feel are more reasonable than those written up in the lab report. Some of the data are unreliable, according to Sweeney, and can be ignored if the students choose. "Again, they have to evaluate. If they don't, they can still come up with a result, but their results won't be as valid as the results derived by the more insightful students." Sweeney has tried this last technique only once, but it worked so well, he says he'll continue using it.

This exam, like most of Sweeney's, are completed at home, generally within 5 days. Students can use class notes, answer keys, texts, reference books, research articles, and whatever else they find helpful—with the exception of other people.

In an effort to use outcome-based grading, students were allowed to try to answer missed exam questions by tackling new questions on the same topic in successive exams. Course grades were based on the number of topics completed by the end of the term. But it was "too much work" for Sweeney to "make up all the new questions," so he hasn't used it again; however, students liked this option a great deal.

Several colleagues responded with enthusiasm to Sweeney's methods after he made a presentation on his lab efforts at an American Chemical Society meeting.

OPEN-BOOK EXAMS

Harry Ungar, Department of Chemistry, Cabrillo Community College, Aptos, CA 95003; TEL: (408) 479-5059; FAX: (408) 464-7778; E-MAIL: haunger@cabrillo.cc.ca.us.

Courses Taught

- Organic Chemistry

- Organic Chemistry (for nursing students)
- Preparatory Chemistry

Description of Examination Innovation

Harry Ungar uses open-book exams for most of the year in his two-semester organic chemistry course taken by premedical and other students in health-related fields, as well as by chemistry and biology majors. For the first half of the first semester, he uses conventional closed-book exams because he wants the students to memorize a large amount of material, particularly nomenclature. After that, all testing is open book. Students are allowed to bring any written materials with them.

Ungar's intention is to break away in the second half of the first semester from memorization. From student feedback, which he routinely solicits, he has gained the impression that memorization is their dominant mode of study, especially before exams. In place of memorization, Ungar encourages his students to rewrite their notes by correlating them with their textbook, arranging and constructing them in a coherent manner, and also to annotate their textbooks in such a way that they can find information quickly.

Ungar's exams are lengthy, so a student who spends too much time looking up material related to one question will not have time to finish all the others. In constructing their notes, he advises students to do so not only on a chapter-by-chapter basis, but also on the basis of themes that run through many chapters, themes such as structure and nomenclature, major mechanisms, stereochemistry, analysis (including spectroscopy), and synthesis. His aim is to have his students integrate the material by mapping out the course content in their own way or together with a group.

Because of the open-book format, Ungar feels free to write exams that focus on the concepts, theories, and major themes of organic chemistry, along with problem solving. "In short," he reports, "I am trying to get them to start thinking like chemists."

Has he succeeded? He says he doesn't really know, but feels confident because of the following outcomes:

1. When he administered the ACS organic exam, a standard, closed-book, multiple-choice exam, students were told that the exam would not count toward their grades, and that they were not to make any special preparation for it. (Nor was his teaching content oriented toward the topics on this standard exam.) Still, the average was exactly the same as the national norms.

2. Many students have gone on from his course to 4-year colleges and medical schools, and have done quite well. None have ever found themselves in any way ill-prepared for the courses that build on organic chemistry—chemistry, biochemistry, or biology courses.

3. A good percentage of Ungar's students say they enjoy his course. In striving to understand the more difficult concepts, they are challenged and intrigued. They appear to "celebrate," Ungar says, when they arrive at a deeper understanding of the mechanism of a reaction in ways that straightforward problem solving doesn't usually deliver.

LONG, COMPLEX QUESTIONS BASED ON DATA

Gerald R. Van Hecke, Department of Chemistry, Harvey Mudd College, 301 East 12th Street, Claremont, CA 91711-5990; TEL: (909) 607-3935; FAX: (909) 607-7577; E-MAIL: gerald_vanhecke@hmc.edu.

Courses Taught

- Physical Chemistry
- Group Theory, Quantum Mechanics, and Spectroscopy
- Classical and Statistical Thermodynamics

Description of Examination Innovation

In Gerald Van Hecke's physical chemistry course, a single question is posed that can take up to 4 hours to answer, requiring between 15 and 20 pages of notes and calculations. Students may use notes, homework, the textbook, and other resources for the exam. The format was developed by Van Hecke and extended by his colleague Kerry Karukstis, and is being used in the various physical chemistry courses they both teach.

In the question, various pieces of data are provided. Students are told that "a solution to the question is possible using the provided data, and only the provided data." The examination, as the instructors see it, is in no small part a puzzle, the solution of which involves at least two steps: (1) to see what information a given piece of data provides, and (2) to demonstrate one's ability to use that information to build toward the solution of the primary problem. (Copies of the single-question examinations are available from Van Hecke.)

As part of the instructions, students are admonished to indicate what they are doing as they proceed. The problem is graded, they are told, in terms of the "correct pieces" that would lead to the final answer. "While numerical results are expected, " the instructors write, "the greatest credit will be given for the correct approach, formula, and so forth." In addition to using notes and the textbook, students may use their corrected homework assignments. (Van Hecke returns homework with hints until they are correct.)

"This type of question was chosen to bring all the concepts learned in the course to bear on a real-life question," says Van Hecke. "All scientists (and engineers) constantly must solve problems that often draw on data from diverse sources." Students from 20 years ago have told Van Hecke they still remember his final as challenging and worthy of their talents.

TWO METHODS OF COOPERATIVE ESSAY-WRITING USING RESEARCH

Eugene J. Volker, Department of Chemistry, Shepherd College, Shepherdstown, WV 25443; TEL: (304) 876-5285; FAX: (304) 876-3101; E-MAIL: evolker@shepherd.wvnet.edu.

Courses Taught

- Organic Chemistry
- College Chemistry (for nurses)

Description of Examination Innovation

Two distinctly different in-class examination methods have been developed by Eugene Volker. In the period 1977–1987, he gave essay-type examinations on previously announced research topics in organic chemistry and a number of other chemistry classes, accounting for half of the student's grade. (The other half came from multiple-choice tests on conventional material taught in the course.) The features of this highly unorthodox method in the physical sciences,[6] taking as an example the organic chemistry course, can be summarized as follows:

- For each of the four 2-hour examinations in the course, a list of 10–15 research topics was given to the students. Each topic contained questions or assignments that had to be answered.
- Students researched the topics individually or in groups in the library and wrote draft essays that could be discussed with the instructor.
- The examination consisted of four or five topics selected from the ones that were handed out. No reference materials, especially no draft essays, were allowed in the room, the students having to write their test from memory.

The advantages of the system were that both theoretical ("Discuss the nature of carbocations...") and practical ("Describe the manufacture of sucrose

---

[6] Described in the *Journal of College Science Teaching* 10(1980): 102.

from its natural sources...") questions could be included, driving home the point that organic chemistry is involved in many facets of everyday life. Another strong point was that students became familiar with the resources of the library. Furthermore, they developed well-functioning study groups and interacted frequently with the instructor. Student evaluations were generally positive during the entire project. A disadvantage is that large amounts of the instructor's time were taken up by student conferences and, especially, by the grading of the essays.

A second approach, which grew out of the first one but has a smaller effect on the student's grade (it is worth 20 percent, with only 4 percent of the material appearing on the in-class examinations), was used from 1992–1995. This method, which has undergone refinement every year, is an attempt to tie chemistry to the "writing across the curriculum" effort. Currently, it involves college chemistry (for nurses) and organic chemistry. The main features are as follows:

- A list of eight research topics per semester, similar in structure to the ones mentioned above, is given to the students. The topics emphasize applications of chemistry ("Give a detailed description of how C-14 is used to date archeological objects...") and research work at the frontiers of chemistry ("How can antibodies control the stereochemistry of the Diels–Alder reaction...").
- As before, the students use the resources of the library (which nowadays means an increasing amount of computerized database searching) to collect material, either individually or in small groups.
- In contrast to the previously described method, they write polished essays and hand these in for a grade. The in-class exams have only a limited number of bonus, multiple-choice questions related to these papers.
- The essays are graded for scientific accuracy, organization, clarity of exposition, and freedom from major syntactic and spelling errors.

Student response, as assessed by detailed questionnaires, is cautiously positive. The most favorable features cited were the ability to understand applications of chemistry and to find out about new developments in research, with lesser support indicated for becoming familiar with information retrieval and technical writing. On the negative side, many students consider the extra workload as burdensome and claim that it takes time away from the study of traditional "core" material. More recently, the instructor has allowed students to hand in cooperatively authored papers that must be 50 percent longer than standard individual essays. The students must sign a statement to the effect that both students have contributed almost equally to the work. In 1996, about 20 percent of all papers were jointly authored.

EXAM REWORKS

Thomas D. Walsh, Department of Chemistry, University of North Carolina-Charlotte, 9201 University City Blvd., Charlotte, NC 28223; TEL: (704) 547-4400; FAX: (704) 547-3151; E-MAIL: tdwalsh@unccvm.uncc.edu.

Courses Taught

* Organic Chemistry I and II

Description of Examination Innovation

Since the fall of 1986, Thomas Walsh has used exam reworks to help his organic chemistry students learn more chemistry.

A few days after his class has taken an exam, he gives a fresh copy of the test to each student. Students have two opportunities outside class to rework the questions for which they did not receive full credit. They may work together, but since the purpose of the rework is to identify those concepts not yet understood, this is best done by working alone.

At the end of the first attempt, Walsh collects the papers and checks them again, and then returns them with the number still incorrect for a final try. If a student reworks all the questions correctly by the second try, he rewards the student by adding two bonus points to his/her exam total. If only one small error remains at that time, one point may be given. Otherwise, no points are gained by the student.

Walsh admits a two-point bonus is not a lot, but nevertheless, the rework option is a good motivator, in that it gives students the rare experience of working on a set of problems until they solve all of them correctly.

"Too often, college students get 7 or 8 out of 10 questions right on an exam and then move on, without learning from their mistakes," he says.

How do students do on reworks? Walsh says it depends on the difficulty of the material and the students' available time. Early in the term, more than half succeed in getting all the problems correct and receive the two extra points. But by the third exam, when students are more pressed for time and the material is more challenging, few are so successful. (There isn't time for a rework option after the final exam.) He used to give students three tries, but when students' preparation for the next exam suffered, he curtailed that practice.

A problem with the reworks, according to Walsh, is that the some students rely too much on members of their study groups to help them. "Working in study groups sometimes gets out of control, and then too few students are solving the problems themselves," says Walsh. For one term in 1987, he gave four bonus points to students who succeeded in a rework, but then reduced this to two points because he was uncomfortable giving that many points for work not necessarily done on their own.

The purpose of the rework is to see if students understand the material under optimal conditions. Some concepts resist understanding even then. "But my purpose is not always *their* purpose," says Walsh. Although he tells his class that even the student who succeeds in all three rework options won't see a difference in the course grade, Walsh believes they have a sort of "mystical faith" in the power of those two points.

Whatever the motivation, Walsh says all his students try the reworks, which tells him that, at the least, they're learning the difference between not remembering the chemistry and never understanding it in the first place.

Student reaction to the rework option is "enthusiastically positive," says Walsh. His colleagues, however, are skeptical of the innovation because of the extra grading time they think it takes. But Walsh says it isn't as bad as it sounds. Partial credit is given on the exam but not in the rework process. "I just indicate which problems still need work," says Walsh. "I keep doing that until all the students' answers are correct, or the two tries are used, and a final exam grade must be given."

VARIANT ON EXAM RETAKE

Tom Werner, Department of Chemistry, Union College, Schenectady, NY 12308; TEL: (518) 388-6335; FAX: (518) 388-6795; E-MAIL: werner@gar.union.edu.

Courses Taught

- Introductory Chemistry
- Analytical Chemistry

Description of Examination Innovation

Tom Werner returns the first of two hour exams in his analytical chemistry course graded but uncorrected. Students are told they have one week to figure out what they did wrong. They can use any source, including one another. The following week, he collects their original exams, gives them a clean copy of the exam, and allows them to make any changes or additions. Then he regrades their exams based on their changes. The final grade for the exam is calculated as the average between the two grades. Students can thus recover one of two points missed on the original.

Most students recover near the maximum number of points on the retake. Good students feel the method is fair, because they still benefit significantly from doing well the first time, and they have little or no preparation to do for the retake.

Werner calls this a "redemption" exam and finds it to be a much more effective pedagogical tool than simply telling students what they missed. He says students who don't know about his method are relieved when he announces it upon returning the first exam. Tension is released, and students particularly appreciate being given a second chance early in the term. This way they don't have "such a big hole to climb out of to get a reasonable grade," says Werner. In this sense, he feels the method improves student morale. At least one of his colleagues uses the approach occasionally as well.

UNTIMED COMBINATION GROUP/INDIVIDUAL EXAMS, PROBLEM-BASED LEARNING

Harold B. White, Department of Chemistry and Biochemistry, University of Delaware, Newark, DE 19716; TEL: (302) 831-2908; FAX: (302) 831-6335; E-MAIL: halwhite@brahms.udel.edu.

Courses Taught

- Introduction to Biochemistry

Description of Examination Innovation

In Introduction to Biochemistry, Harold White has used a problem-based learning model that requires students to read classic articles in the field. The articles serve as problems for the students to analyze and understand. In place of lecturing, White organizes his class of sophomore biochemistry majors into small groups that must define, through discussion, what they don't understand about each article and then find out what they need to know in the library or elsewhere. During class, he moves among the groups, giving pointers and encouraging further analysis.

Although White had used research articles before in this course,[7] he had always suspected that students "simply skimmed an article, perhaps looked up a few words, got the 'gist' of it, and thought they understood it." Now, with the course restructured around a problem-based learning model, he observes his students in class talking about the articles and what they mean in a way he

---

[7] See H. B. White, "Introduction to Biochemistry: A Different Approach," *Biochemical Education* 20(1992): 22–23; and, for a more recent description of how the course is run, see H. B. White, "Addressing Content in Problem-Based Courses: The Learning Issue Matrix," *Biochemical Education* 24(1996): 41–45.

associates with faculty discussions. Some of the questions they ask are so good that he has not thought about them before.

When it comes to examining his students' work, White runs into a few problems. Some students mind that he assigns 20 percent of their grade to attendance, preparation, participation, and attitude. The fact that their grades are determined in part by the work of their groups has also been a sticking point. Others feel unsure about what is expected: They prefer having the teacher lecture about each article and tell them what is important.

For the midterm, White chooses an article students haven't read before, assigns it in advance of the examination, and asks pertinent questions about its content, methodology, and significance. Part I of the examination (75 percent of the grade) is done individually and handed in. Part II takes the last and most difficult question from Part I and assigns it for group analysis. After discussion, the group prepares and turns in an answer (25 percent of the grade). If a group cannot come to consensus, group members may hand in separate answers for an individual grade.

Even the part of the examination taken individually is different from a traditional exam. Having stressed "learning issues"—"knowing what you don't yet know"—during the course, White asks (for 20 percent of their grade) that students make a list of their "remaining learning issues," ranked in order of importance and justified with respect to their importance. The midterm is taken at students' own pace.

Grading essay answers turns out to be painful. "There is no way to evaluate answers objectively," says White. Some students seem to understand a concept but can't communicate that understanding. Some take paragraphs to explain what others can say in a single sentence. Some deal with peripheral misinformation, and the instructor has had more than a few "beautifully written wrong answers" to deal with. "How can you explain to a student why the 'quality' of his answer was poor?" White asks.

Even though the instructor now feels more comfortable assessing quality by assigning individual grades for Part I, he is still seeking ways to appropriately measure the contribution of group performance in Part II. Despite these vexing issues, White is convinced of the educational value of the problem-based approach and the examination mode he has adopted. He expects to continue using both for the forseeable future.

CHAPTER 4

PHYSICS

INCORPORATING GRAPH INTERPRETATION ON TESTS

Robert J. Beichner, Department of Physics, North Carolina State University, Raleigh, NC 29695; TEL: (919) 515-7226; FAX: (919) 515-6538; E-MAIL: beichner@ncsu.edu.

Courses Taught

- Conceptual Physics
- Engineering Physics
- Advanced Physics Lab

Description of Examination Innovation

"The ability to comfortably work with graphs," writes Robert Beichner, quoting a 1933–34 American Physical Society Committee on Tests Report,[1] "is a basic skill of the scientist." Concerned that interpretation of graphs is not sufficiently *assessed* in kinematics, Beichner set about devising a "single, consistent assessment instrument" that would be helpful both to researchers trying to compare results from several studies and to instructors assessing their success at teaching students' how to construct and use scientific representations.

The 21 multiple-choice questions that resulted from his efforts have been carefully validated in terms of accuracy in measuring graph-interpretation skills and stability (reliability) over large numbers of test takers. Beichner took pains to ensure that only kinematics graph-interpretation skills were being measured by the test items. For example, as Beichner writes, "an item asking a student to

---

[1] Robert J. Beichner, "Testing Student Interpretation of Kinematics Graphs," *American Journal of Physics* 62, no. 8(August 1994): 750–762. This entry is largely derived from that article, with the permission of the author.

select the graph which correctly describes the vertical component of the velocity of a ball tossed into the air would be inappropriate, since it tests knowledge of projectile motion."

Items and distracters were deliberately written so as to attract students holding previously reported graphing difficulties. Beichner accomplished this by asking open-ended questions of a group of students and then using the most frequently made mistakes as distracters.

Beichner's motives in performing the experiment are summarized in his "Implications for Instruction":

> Students need to understand graphs before they can be used as a language for instruction.... [I]nstruction incorporating these graphs must include thorough explanations of all the information each one relates.... [T]eachers must choose their own words carefully—for example, the word "change" does not automatically signify "find a slope"—and be alert for similar mistakes when students are involved in discussions amongst themselves or with the instructor.
>
> Students should be asked to translate from motion events to kinematics graphs and back again...[and] to go back and forth between the different kinematics graphs, inferring the shape of one from another... (p. 755)

Of as much interest to us (and to our readers) are the test items themselves and so, with the permission of the researcher, we have reproduced them on pages 107–109. In addition, further information and the latest version of the test are available on the World Wide Web at http://www2.ncsu.edu/ncsu/pams/physics/Physics_Ed/ TUGK.html.

INDIVIDUALIZED PROJECT AND COMPREHENSIVE EXAMS

Herbert Bernstein, Division of Natural Sciences, Hampshire College, Amherst, MA 01002; TEL: (413) 582-5573; FAX: (413) 582-5448; E-MAIL: hbernstein@hampshire.edu.

Courses Taught

- Quantum Mechanics
- General Relativity
- Physics I, II, and III
- Biophysics
- Electronics for People
- Book seminars on "E & M," "Mechanics," "Relativity," and "Quantum Theory"

Description of Examination Innovation

Hampshire College is an innovative, experimental, liberal arts college of 1100 students in the Five-College Consortium (together with Amherst, Smith, and

❶ Acceleration versus time graphs for five objects are shown below. All axes have the same scale. Which object had the greatest change in velocity during the interval?

❷ When is the acceleration the most negative?

(A) R to T
(B) T to V
(C) V
(D) X
(E) X to Z

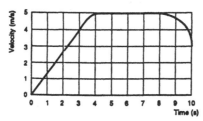

❸ To the right is a graph of an object's motion. Which sentence is the best interpretation?

(A) The object is moving with a constant, non-zero acceleration.
(B) The object does not move.
(C) The object is moving with a uniformly increasing velocity.
(D) The object is moving at a constant velocity.
(E) The object is moving with a uniformly increasing acceleration.

❹ An elevator moves from the basement to the tenth floor of a building. The mass of the elevator is 1000 kg and it moves as shown in the velocity-time graph below. How far does it move during the first three seconds of motion?

(A) 0.75 m
(B) 1.33 m
(C) 4.0 m
(D) 6.0 m
(E) 12.0 m

⑤ The velocity at the 2 second point is:

(A) 0.4 m/s
(B) 2.0 m/s
(C) 2.5 m/s
(D) 5.0 m/s
(E) 10.0 m/s

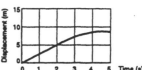

⑥ This graph shows velocity as a function of time for a car of mass 1.5 x 10³ kg. What was the acceleration at the end of 90 s ?

(A) 0.22 m/s²
(B) 0.33 m/s²
(C) 1.0 m/s²
(D) 9.8 m/s²
(E) 20 m/s²

⑦ The motion of an object traveling in a straight line is represented by the following graph. At time = 65 s, the magnitude of the instantaneous acceleration of the object was most nearly:

(A) 1 m/s²
(B) 2 m/s²
(C) +9.8 m/s²
(D) +30 m/s²
(E) +34 m/s²

Ⓗ Here is a graph of an object's motion. Which sentence is a correct interpretation?

(A) The object rolls along a flat surface. Then it rolls forward down a hill, and then finally stops.

(B) The object doesn't move at first. Then it rolls forward down a hill and finally stops.

(C) The object is moving at a constant velocity. Then it slows down and stops.

(D) The object doesn't move at first. Then it moves backwards and then finally stops.

(E) The object moves along a flat area, moves backwards down a hill, and then it keeps moving.

⑨ An object starts from rest and undergoes a positive, constant acceleration for ten seconds. It then continues on with constant velocity. Which of the following graphs correctly describes this situation?

①⓪ Five objects move according to the following acceleration versus time graphs. Which has the smallest change in velocity during the three second interval?

①① The following is a displacement-time graph for an object during a 5 s time interval.

Which one of the following graphs of velocity versus time would best represent the object's motion during the same time interval?

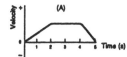 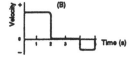 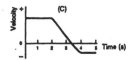

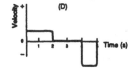 

Mount Holyoke colleges, and the University of Massachusetts). But Hampshire has developed its own unique inquiry-oriented introductory science-education program. To fulfill their science requirements, all students take inquiry courses that focus on particular topics, leading to individual research projects on topics of the students' choosing. After the course work, projects are completed in tutorial fashion, with individual faculty meeting with the students and guiding them through revisions until the projects are of sufficient quality. This so-called "examination," the primary innovation, incorporates research into the very beginning of each student's education. In general, course examinations are deemphasized in favor of individualized project "examinations" throughout the curriculum.

As is typical in the natural sciences at Hampshire, homework is assigned in the previously listed courses, but individualized projects are what teach students how *real* science is actually done. These individualized assignments usually involve a written paper and an oral presentation. For "Division II" students, who have taken courses at a higher level at the Five-College Consortium and at Hampshire, Herb Bernstein devises a "comprehensive examination" for each of the physics concentrators.

The examination is individualized, in that the instructor consults each student's unique list of courses, projects, and learning activities (including internships) to determine which questions, from a uniform examination (based on the comprehensive examination for a master's degree), will appear on the student's exam. Students have 1 or 2 questions, out of as few as 10 or as many as 16, which provide options (e.g., Do problem 7 or problem 9), but the options themselves are set by the examiner. Students almost always attempt over 80 percent of the assigned problems. This examination is nonstandard in many ways, but it is administered in the standard manner—written, closed-book format.

CONCEPTS ON QUIZZES, MULTIPLE-CHOICE WITH EXPLANATION

Richard Crowe, Department of Physics and Astronomy, University of Hawaii at Hilo, 200 West Kawili Street, Hilo, HI 96720; TEL: (808) 933-3649; FAX: (808) 933-3693; E-MAIL: rcrowe@maxwell.uhh.hawaii.edu.

Courses Taught

- General Astronomy and Lab
- Principles of Astronomy
- Stellar Astrophysics
- Introduction to Modern Physics
- College Physics and Lab
- Advanced Classical Mechanics
- Advanced Modern Physics
- Cosmology

Description of Examination Innovation

In his introductory physics and introductory astronomy courses, Richard Crowe uses the analytic or critical-thinking question. He employs a modified multiple-choice format, which requires students to make a choice, justify their choice, and explain why they rejected the other choices. A correct, even justified, choice lacking an explanation of why the others were rejected qualifies for only half credit.

A typical question asks students to provide a likely explanation for both the absorption lines of ionized helium and the molecular absorption bands of titanium oxide in the photographic spectrum of a star, as observed from the Earth. Distracters include (A) "a hot O star surrounded by a thin shell of 3000-degree gas," (B) "a cool M star surrounded by a thin shell of 30,000-degree gas," and three others. Students must choose the correct answer and then explain why they selected that answer, and why they rejected the other five distracters provided.

The correct answer is a binary star, one an O star, the other an M star. An acceptable explanation for why (A) is incorrect is that if the surface temperature were 30,000 degrees, it would likely heat an outer shell to much more than 3000 degrees. (B) is incorrect because if the shell were hotter, it would generate helium emission lines, not absorption lines.

Crowe has been using this testing system for 5 years. Some students have complained on course evaluations that the questions are too difficult, and Crowe acknowledges that it is hard to get full credit on these questions. Sometimes he uses them as extra-credit questions and always has students working them out in advance in self-organized student groups. Overall, the group format stimulated interest in the material and was effective so long as all members of the group participated. Some of Crowe's colleagues use concept-oriented approaches in their classes, but none use Crowe's specific testing format.

Crowe also administers weekly quizzes that test students solely on conceptual material. These quizzes allow him to reduce the number of term tests and use homework assignments, which lack time pressure, to assess problem-solving skills. Overall, Crowe sums up his philosophy of testing and grading as follows: "I believe that introductory courses should focus on understanding and be concept oriented. Problem-solving (quantitative) skills are important but can be developed later."

GROUP PROJECTS

Martin Hackworth, Department of Physics, Box 8106, Idaho State University, Pocatello, ID 83209; TEL: (208) 236-4439; FAX: (208) 236-4649; E-MAIL: hackmart@physics.isu.edu.

Courses Taught

- Engineering Physics

- Physics Labs
- Astronomy Lab
- Introduction to Math for Physics
- Introduction to Calculators and Computers for Physics Majors

Description of Examination Innovation

In his engineering physics and introduction to calculators and computers for physics courses, Martin Hackworth uses graded group projects. Typical of the questions that groups of two to four students (depending on the task) tackle is the following: "Compare and contrast a bouncing ball and a ball on the end of a spring. Are the motions fundamentally the same or different?"

Group members describe the motion they see and state their beliefs. Members' next task is to graph the displacement, velocity, and acceleration versus time for each system and compare results. A summary of each group's initial ideas, graphs, conclusions, and physical reasoning is turned in for a single grade per group.

Groups are randomly assigned by the instructor for each project. This way, the instructor finds that each group sees a greater contribution from individual students. And because grades are not based solely on these exercises, students who limit their participation in the group work (a minority, in his experience) still have a good portion of their course grade (between 50 and 90 percent, depending on the course) determined by individual effort.

In these settings, Hackworth sees himself as "tour guide" for the students' journey of discovery in physics and scientific thinking. He introduces and discusses the concept, problem, or exercise with students and then allows them to work on the problem (with lab equipment if needed) by themselves. He answers questions only if by answering them he can elicit information from the student as well. He sees himself as verifying fundamental physical principles and helping with technical details.

The group projects are popular with students, according to end-of-course evaluations. Students also seem more comfortable with the grading process, and, because they feel vested in the process, are less likely to complain about grades.

MIDSEMESTER LAB PRACTICAL

Fred Hartline, Department of Physics and Computer Science, Christopher Newport University, Newport News, VA 23606-2008; TEL: (804) 594-7181; FAX: (804) 594-7919; E-MAIL: fhartline@pcs.cnu.edu.

Courses Taught

- Physics 201 and Laboratory

Description of Examination Innovation

In introductory physics labs, Fred Hartline has observed three related problems. First, students often rely on one another—to the point of "parasitism" by weak students. Second, there is never enough lab time to explore and learn about relevant phenomena. And finally, students want their instructor to tell them what to do in lab.

To combat these problems, Hartline and his colleagues at Christopher Newport developed a physics lab practical midsemester exam 6 years ago that they've been using ever since. Stations are set up for each of five different 30-minute activities. Two are experimental explorations (making measurements and limited analysis), two are graphical analyses of provided data sets (one uses MathCAD), and one is a lab-based "fundamental skills and concepts" inventory test. Students work individually and rotate through all five activities in the course of the 3-hour lab.

What makes Hartline's lab practical interesting (and different from the traditional practicum) is that students are doing activities they have never seen before. All activities involve basic principles that must be applied in novel situations using equipment with which students are familiar.

For example, Hartline doesn't cover pendula explicitly in class. But in the lab practical, students will be asked to graph the period of a ring pendulum as a function of the ring diameter. Students are given both a photogate timer and a stopwatch. Hartline wants to know whether they can figure out how to use the equipment to do the necessary measurements. Either piece of equipment can be used, depending on the student's preference, though in most exercises, one choice is superior for determining a value.

"Our goal is to make the exam a valuable learning experience," says Hartline, "and at the same time to diagnose the extent to which our students have learned generalized skills and techniques that will help them solve new real-world problems." Therefore, students take the lab practical exam individually, not in a group.

Today, Hartline and his colleagues provide more detail to students about the exam in advance than they did 5 years ago, mainly because the format is (still) so unusual. The change gives students more to work with and learn, says Hartline. They can review concepts and try to brainstorm approaches with one another before seeing the actual activities.

Many students still feel pressured and rushed during these physics lab practicum, and although for some the pressure wanes as the semester continues, Hartline continues to work on ways to "humanize" the experience to reduce student pressure without diminishing the evaluative function of the exam.

Two years ago, Hartline started using collaborative group exams in a physical science course for preservice teachers. These exams are enthusiastically received by the students, even though the explorations cover new material.

Although each student is responsible for specific parts of the exam, all are reassured by the supportive nature of the collaboration. This approach successfully addresses Hartline's exam objectives (except for the problem with very weak students) and is being considered as a modification for the lab practical.

DAILY QUIZZES, SINGLE-QUESTION MIDTERM, CONCEPTUAL FINAL

Carl S. Helrich, Department of Physics, Goshen College, Goshen, IN 46525; TEL: (219) 535-7302; FAX: (219) 535-7509; E-MAIL: carlsh@goshen.edu.

Courses Taught

- Physics (general education)
- Methods of Mathematical Physics
- Quantum Mechanics
- Analytical Mechanics
- Electricity and Magnetism
- Thermodynamics
- Solid-State Physics
- Science and Religion

Description of Examination Innovation

Carl Helrich gives nearly daily quizzes in most of his physics courses in an effort to keep students current in the class. He got tired of having students who assumed they could put off studying material as complex and densely packed as upper-division physics courses until the days before the exam. His quizzes are intended to cover material from the previous class.

Questions on the 10-minute quizzes are not algorithmic. Rather, the instructor asks for an explanation of a point in the previous lecture or discussion. Even though Helrich realizes that some students still study for the quiz only in the minutes before class, he believes that daily quizzes prepare the student for the next lecture or discussion.

When the instructor first introduced his almost daily quiz system, he continued to give, in addition, three standard exams and a final. In recent years, however, he has replaced the midterm with a single take-home problem or an analysis requiring creative thought. In analytical mechanics, for example, a recent question called for a detailed write-up of a scattering problem, with explanations and any necessary derivations. In quantum mechanics, the single question called for an outline of the ideas of Dirac.

For Helrich's final examination in the last few years, students write a detailed description of certain aspects of the course and then solve one problem from each. In thermodynamics, for example, the instructor might ask for a detailed discussion of the application of thermodynamic stability (minimization of the Gibbs function) to chemical equilibrium, followed by a problem employing these concepts. Such an exam tests understanding of the concepts, not simply memorization of any one formulation. In the grading of such an examination, Helrich employs a protocol that lists the ideas he hopes students will come up with in their discussion and assigns points depending on how many and how well they cover these.

Among the many benefits of his comprehensive new examination strategy, the instructor notes the elimination of the "cold sweat and panic" particularly common to engineering students. In his personal evaluation of his classes, Helrich finds a positive response to the quiz system. As for faculty response, Helrich works in a two-person department and reports that his departmental colleague has not followed suit. Neither have his colleagues in chemistry, although one has begun inquiring about the approach.,

GROUP EXAMS AND RETESTS

Claud H. Sandberg Lacy, Department of Physics, University of Arkansas, Fayetteville, AR 72701; TEL: (501) 575-5928; FAX: (501) 575-4580; E-MAIL: clacy@comp.uark.edu.

Courses Taught

- Astronomy 2003–"Survey of the Universe"
- Astronomy 3053–"Stellar Systems"

Description of Examination Innovation

In the fall of 1994, Claud Sandberg Lacy tried two new practices he thought would encourage student interaction in and outside of class and would eventually lead to improved learning. The first new practice was to organize groups of three to four members who work together throughout the semester and who are, on occasion, asked to respond to discussion questions as a group. In the middle of a lecture/presentation, Lacy will pose questions intended to generate discussion about the topic being covered, such as, "Describe a theory that successfully explains the origin of the moon. The theory should explain the observed lack of iron in the moon, relative to the earth, and the lack of water and other volatiles in rocks brought back from the moon." Then he gives the groups 15 minutes to discuss and answer the question in writing. Each group chooses a "reporter" for the day. Group discussion questions (about 10 per semester) count as 15 percent of the course grade.

Lacy calls his second new practice "retests." Between the completion of an exam (of which there are three during the semester) and the next class, each study group meets outside of class to discuss the test questions and tries to come up with the correct answers. Group members then fill out another answer sheet (computer-gradable) and hand it in.

Students may increase their initial grade on each test by one-third if their second set of answers is correct. A test score of 70 percent, followed by a retest score of 100 percent, would generate an effective score of 80 percent in Lacy's scheme. In a midsemester poll of the class, the two most popular features of Lacy's courses were retests and discussion groups.

Lacy got the idea for these two innovations from discussions with other instructors while participating in the Chancellor's Teaching Mentor program at the University of Arkansas. Because test and retest grading and rostering are handled by computer, the method takes minimal time. The discussion questions take the most time to grade by hand (about 2–3 hours per question to grade and record in a 170-student class), which influences whether Lacy will ask such a question in a particular week.

CHAPTER 5

ASSORTED FIELDS

SUBJECT-RELEVANT CARTOONS ON EXAMS

R. Lynn Bradley, Department of Geography, Belleville Area College, Belleville, IL 62221-5899; TEL: (618) 235-2700 Ext. 406; FAX: (618) 235-1578.

Courses Taught

- Earth Science
- World Regional Geography
- Geography of the United States and Canada

Description of Examination Innovation

For the past 3 or 4 years, R. Lynn Bradley has included cartoons on his exams. Previously, he collected cartoons dealing with the material that he teaches. It occurred to him that including cartoons on exams might help relieve some of the anxiety that accompanies test taking. He finds that the interspersion of cartoons on his exams works well. He's received several favorable comments from students and heard "chuckles" during some exams. One colleague started using cartoons to go with her lectures, which Bradley also does.

WEEKLY QUIZ REWORKS

Peter C. Chang, Department of Civil Engineering, University of Maryland, College Park, MD 20742; TEL: (301) 405-1957; FAX: (301) 405-2585; E-MAIL: pchang@eng.umd.edu.

Courses Taught

- Mechanics of Materials—sophomore-level introductory course

Description of Examination Innovation

Peter Chang's testing innovation was designed to solve two problems in his course: the large amount of material taught, and the fact that many hour exams are given at the same time in the semester. One remedy was to put less value on the hour exams and provide incentives to students to keep up with the material being taught. Chang now divides the grade into three areas of approximately equal weight: (1) hour exams, (2) homework problems and weekly quizzes, and (3) design studio and reports. The homework problems are due *before* the material is covered in class. This, according to Chang, was the most important change and results in more lively classes, since students now come to class with the work done and questions to pose. The weekly quizzes are on material just covered during that week, and quizzes can be retaken within a week if the student does poorly the first time. Altogether, the changes have redirected emphasis from test performance to learning.

In the future, Chang intends to modify his hour exams to rely less on textbook problems and instead more closely resemble professional engineering examinations. He regrets, however, that "a large percentage of students see this exam as their ultimate goal instead of learning the material." Significantly, he says that because students are so "conditioned" by the traditional grading system, many do not believe (however many times he tells them) that the hour exams together constitute only 30 percent of their grade.

Chang has been using the new approach for six semesters but has had to make adjustments to handle unanticipated problems. One problem was that students liked being able to retake quizzes but abused the system by retaking them until they got the score they wanted. Now Chang limits quiz retakes to five (total) over the semester. Another problem was that although students were willing to try to solve problems before the material was taught, their homework grades took a plunge. To remedy this, homework can be reworked and turned in by the end of the week, before the assignment is graded.

The biggest unsolved problem is having to give students a single grade at the end of the semester—because a single grade "cannot adequately describe the students' ability." Chang wants to see a grading system that distinguishes creativity, written and oral presentations, test performance, and ability to work with others.

Colleagues have not adopted Chang's grading system, he reports, because it requires so much extra work on the instructor's part. Unfortunately, says Chang, "without a culture of change throughout academic institutions, individual efforts are not likely to make much impact."

PAIRS TESTING

Don L. Dekker, Department of Mechanical Engineering, Rose–Hulman Institute of Technology, Terre Haute, IN 47803-3999; TEL: (812) 877-8327; FAX: (812) 877-3198; E-MAIL: don_dekker@rose-hulman.edu.
James E. Stice, Professor Emeritus, Department of Chemical Engineering, University of Texas–Austin, Austin, TX 78712-1062; TEL: (512) 471-5238; FAX: (512) 471-7060; E-MAIL: stice@che.utexas.edu.

Courses Taught

- Thermal Design

Description of Examination Innovation

For nearly 15 years, some mathematics, science, and engineering educators have been using an alternative technique for practicing problem solving. Popularized as the "Whimbey pairs method," the technique involves thinking aloud while problem solving.[1] As a pair, one student is the "problem solver" and the other the "listener." The problem solver reads the problem aloud and then continues to talk aloud as he/she proceeds to try to solve it. The listener listens, asking for clarification whenever the problem solver says anything the listener doesn't understand. This technique greatly expands students' awareness of how others solve problems. One student was amazed when his roommate solved a problem using a method totally different from the one he would have used.

After using pairs–problem solving in his courses in mechanical engineering, it appeared inappropriate to Don Dekker to return to single-answer problems on the exams. So, 10 years ago, he expanded pairs–problem solving into pairs testing. He tried a pairs take-home test and a pairs take-home final exam, and was quite pleased by the results. Solutions to many types of open-ended and design problems are vastly improved by discussion and interaction among students.

Jim Stice, at the University of Texas–Austin, picked up on pairs testing and gives pairs tests in class at night, because they require more time to complete. About a week before a scheduled examination, students are told that they will work as partners on the exam and are urged to give some thought to choosing their partner. At the next class, a sheet of paper is passed around on which they list their names and the name of their partner. When the exam is given, students without a partner are arbitrarily assigned one.

---

[1] Described in Jack Lochhead and Art Whimbey, "Teaching Analytical Reasoning through Thinking Aloud Pair Problem Solving," in J. E. Stice, ed., *Developing Critical Thinking and Problem-Solving Abilities: New Directions for Teaching and Learning,* No. 30. (San Francisco: Jossey Bass, 1987), pp. 73–92.

Students work together during the exam but hand in only one paper. The instructor finds it necessary to allow 50 percent more time for the exam and to select a room that has twice as many seats as there are students so that the groups can spread out.

There are several advantages to pairs testing, Dekker and Stice observe. First, students check each other's work, so fewer careless errors sneak by them. Thus, their papers are easier to grade. Also, the students either get good answers, or they aren't able to do much with a problem, which also aids in grading. Finally, there are only half as many papers to read, which also simplifies grading and makes the method appealing to the instructors.

Students like the method, too. They see it as a rational way to evaluate their competence, much closer to the way they will work when they graduate. Also, they receive appreciably better exam scores—10 to 15 percent better. Exam apprehension is significantly less than when students work alone, and there are indications that motivation is enhanced. Success seems to breed a desire to learn more. Finally, and from the instructor's point of view a real plus, cheating disappears.

On the downside, some pairs have difficulty working together. Some mismatches of personalities occur—too much assertiveness on the part of the one, too much acquiescence on the part of the other; or inability to compromise. Then there are problems of communicating. Also, some students really prefer to work alone. Among the self-confident is an aversion to spending the extra time it takes to go through the problems with someone else. In some cases, a weak or lazy student can let his/her partner do more or all of the work.

The instructor doesn't solve these problems but lets the students figure out how to do so. "After all, in the professional world," says Dekker, "we often cannot choose the colleagues with whom we have to work."

When students insist on working alone, the instructor allows this. (In one instance, two of the loners made the highest grades in the course; in another, the loner made the lowest grade.) In those few cases in which students ride the coattails of an abler friend, the cure is not to make all exams pairs-tested, but to give some conventionally. In one course Stice taught, four hour-long exams and a final were given with pairs testing used on the first and the third hour exams. It was easy to discriminate between partners' abilities from the results of the tests they took individually. Having students work with different partners on successive tests also helps.

Finally, the instructors find that in addition to doing well (and perhaps linked to their higher grades), students are more enthusiastic about the courses involving pairs testing. Students' written responses to the courses are available from the authors.

Dekker and Stice continue to use pairs testing in classes, and in addition, sometimes have a design group of three to four students take on a project as their final exam. They know of no other professors doing this besides themselves.

DESIGN PROBLEM

Karen Den Braven, Department of Mechanical Engineering, University of Idaho, Moscow, ID 83843; TEL: (208) 885-7655; FAX: (208) 885-9031; E-MAIL: kdenb@uidaho.edu.

Courses Taught

- Mechanical Engineering 443
- Solar Energy Engineering and Design (senior elective)

Description of Examination Innovation

In place of a typical final exam, students in Karen Den Braven's solar engineering course are allowed to do a practical design for and/or build solar cookers. They are given price and performance criteria, which range from a $1 hot-dog cooker to a $25 cooker (common in Third World countries) that will cook an entire meal. They are allowed to work in groups. Those who choose not to build the object have to calculate the expected performance of their design. Built or not built, all students must submit a report that includes directions on how to build the cooker—instructions that have to be understandable to a 10-year-old. The cookers created by engineering students have been used numerous times in deomonstrations for local school children.

Den Braven has noticed in her years of teaching that engineering students typically divide into those who are analytically strong and those who prefer practical, hands-on assignments. She feels her final project enables students to work within their strengths. The project covers all the basic theoretical aspects of the course, rewards students for using their imagination, allows for some fun, and reinforces learning by having students "teach"—that is, write a construction manual for kids. Several of Den Braven's students even involved their own children in design, construction, or testing of the cookers. Since the class is taught only once every 2 years, Den Braven has so far only taught it this way once, but she intends to do something similar—probably a solar food dryer—when she teaches the class again.

Den Braven says several of her colleagues in mechanical engineering use similar "design-type" final projects, especially in junior- and senior-level courses. The assessment format could definitely be used in science courses, she believes, but it is particularly applicable to engineering, given the problem-solving emphasis of the curriculum.

GROUP PROJECT/PRESENTATION, OPEN-BOOK/UNTIMED EXAM, ABSOLUTE GRADING

Albert K. Henning, Thayer School of Engineering, Dartmouth College, 8000 Cummings Hall, Hanover, NH 03755-8000; TEL: (603) 646-3671; FAX: (603) 646-2230;
E-MAIL: al.henning@dartmouth.edu.

Courses Taught

- Engineering Sciences 22-Systems

Description of Examination Innovation

Engineering Sciences 22-Systems is a broad-based course covering any system that can be described by the mathematics of ordinary differential equations—mechanical, electrical, thermal, fluid, chemical, biological, and so forth. The mathematics provides the foundation, but the instructor is really trying to engage students in the process of seeing and experiencing the world in terms of models and describing the models' behavior, mathematically and physically.

Ultimately, in final projects, students are asked to compare their model predictions with real observations. Toward the end of the course, small groups of usually three to four students analyze systems of their own choice, taking measurements, creating models, comparing the two, and making a presentation to the entire class. Projects have included analyses of homemade bobsled runs, the dynamics of jello, the behavior of a French horn, economic systems modeled as electrical systems, cross-country skiing, ski jumping, pole vaulting, and hot tubs. The instructor likes the projects, he says, not only because they engage the students in a hands-on activity, but also because they give the entire class an experience with many different systems and unusual analytic techniques, all in a relatively short period of time.

Classes are· small: roughly 35 to 40 students, which makes certain assignments possible that would be too time-consuming to do in a larger class. For example, instead of weekly homework assignments that are collected and graded, for the past 3 years, the instructor has been requiring students to do their problem sets in a personal journal, where they solve and reflect upon their work. This makes it possible for the instructor to lay out pedagogical goals and frameworks for those questions, issues that previously would have been overlooked by the students. Some problems are "word problems," some are open-ended, but all are designed to "stretch students' intuition and push students to their limits, so that performance on more traditional exam problems will feel less difficult," says the instructor. Teaching assistants and the instructor are available at all times to assist.

Only 40 to 50 percent of Henning's students are mature enough to take responsibility for keeping the journals. Because homework problems are not "due" each week, fewer students show up for regularly scheduled problem sessions. As a result, Henning is considering a modification of the journal method. In the future, students may choose, via a written contract, either the journal or a more traditional method of weekly, due-on-time, graded homework assignments.

Henning has also changed his in-class exams. The first exam is an oral exam, which takes place one-on-one for 30 minutes in the instructor's office. The instructor demonstrates a simple system and asks the student to predict the response of the system and verify that prediction using the analytical skills and problem-solving methodology learned in class. The systems may be as varied as a putted golf ball, a blender, or a crosscut saw.

Students are at first fearful of their oral exam, though word is getting around, Henning says, that it is not so bad. Henning's goals are explicit: to get to know each student as an individual, and to be able to gauge his/her strengths and weaknesses in ways that a written exam cannot assess. Henning feels that only through the immediacy of an oral exam can he get at the assumptions (conscious and unconscious) about engineering that students carry around with them, and make them think on their feet and speak, instead of hiding behind a notebook. Since the oral is their first exam, it permits the instructor to identify early on those students who, for emotional and/or intellectual reasons, are at risk in the course.

The second midterm and the final exam are written. There is no time limit for finishing these exams, but students must take them in a single seating. The exams are open book and open notes since, as the instructor tells his students, "This is the way the real world works in engineering." Copies of all old exams, along with problem solutions, are available to students in the library so that they can be acquainted with the "style" of the examination before they see their first. Multiple ways of solution are valued equally by the instructor.

Henning does not grade on a curve. Each portion of the course (journals, labs, written exams, oral exam, project presentation) contributes a certain fraction toward the total grade. But the grade scale (based on the cumulative performance of past years of the class) is set at the start of the course: A certain point score receives a certain grade. Students are challenged to do better than the previous years' performance. They're promised "all As" if they do.

Despite all this, in course evaluations, approximately half of the students say they still need the structure of assigned homework with specific due dates.

EXAM CONTENT SPELLED OUT IN ADVANCE, GROUP QUIZZES

Lueny Morell de Ramirez, Department of Chemical Engineering, University of Puerto Rico, P.O. Box 5000 College Station, Mayaguez, PR 00681-5000; TEL: (787) 832-4040 Ext. 2473 or 2568; FAX: (787) 265-3818; E-MAIL: lueny@dediego.upr.clu.edu.

Courses Taught

- Mass and Energy Balances
- Thermodynamics
- Heat Transfer
- Plastics Technology
- Chemistry Engineering Laboratory

Description of Examination Innovation

Lueny Morell makes it a practice to let her students know the objectives of the test before the test: what they are supposed to know, and what they are supposed to be able to do with that knowledge. For example, prior to a mass and energy balances exam, she will say:

> You are supposed to know the following material and perform the following tasks: convert from one system of units to another; understand what are any process variables; change from mass flow rate to molar flow rate to volumetric flow rate; identify type of system (closed, open, steady state, etc.); find the stoichiometry, the limiting/excess reactants, degree of conversion, product distribution for a given chemical reaction.

Of course, students can always look at the syllabus and find these objectives embedded in it, but telling students beforehand precisely what is expected of them encourages more systematic study. They can test themselves as they study.

The instructor thinks of herself as a "first-grade teacher" (which in some ways she is, since this course is the first chemical engineering course taken by students). She continuously reminds students of what to look for, saying things such as "What do we do next?" "What kind of system is this one?" In the class prior to an exam, she has them answer questions like these in groups. The first group that answers correctly gets bonus points on the exam.

The instructor makes old exams available to students. Old exams, she finds, alleviate test anxiety. A typical attrition rate of 60 percent in this course makes exam taking very stressful for students. Problems are selected for each test by the cohort of course instructors who share the course.

Another of Morell's innovations involves team quizzes. Every 2 weeks or so, the instructor prepares a quiz to be solved in teams formed during the first week of class. She tells students that the purpose of the quiz is for them to find out how well they know the material. They are given 25 to 30 minutes to solve the problem in groups. Once the groups are finished, the class solves the problem together at the blackboard. Then, they exchange papers and grade each other's quizzes. Each group decides the distribution of points. Sometimes, the instructor finds they are "tougher" graders than she.

Classes involving group learning are "quite alive," says Morell. "Students may stand up, walk about, consult with me or other groups, or go to the board to argue a point." It takes a while to switch to this teaching–examining technique, but she finds it is worthwhile. Students agree, because during presemester registration, her sections are the first to fill. Certainly, the fact that 80 percent of her students pass the mass and energy balances course with an A, B, or C, compared to the average 50 percent attrition rate in other sections (using the same standardized tests in all sections) hasn't escaped their attention.

PROBLEM-SOLVING THROUGH INTEGRATION OF CONCEPTS

Donald Steila, late of University of North Carolina at Charlotte.

Courses Taught

- Soil Science
- Dynamic Meteorology
- Synoptic Meteorology
- Introductory Physical Geography

Description of Examination Innovation

Donald Steila found in his teaching that long (80–100 items) multiple-choice examinations dehumanized and generally encouraged memorization of isolated rather than integrated concepts. Readily admitting he had no statistical evidence, he suspected that if, out of a 100-item sample of multiple-choice test questions, 10–20 were randomly selected and used for an exam, the resultant grades would not be impacted.

"If we believe in statistical sampling, why not apply it to the learning environment?" he wondered. "Like many young professors today, I developed examinations that mirror those given by my mentors. After two decades of using the multiple-choice exam, which I never felt comfortable about, I switched."

Steila's approach, in his 200+ student introductory physical geography course was to provide 10–15 problem-solving questions in a mathematical and short-answer format that integrated multiple concepts.

A typical question began with a situation: "A stable air parcel at sea level is forced to lift over a 10,000-meter mountain range. Its initial temperature and dew-point temperature are 30 degrees C and 10 degrees C, respectively." In the multipart question that followed, students had to determine the altitude, temperature, and dew-point temperature at the lifting condensation level; calculate the temperature and dew-point temperature along the ridge of the mountain range; and determine the final temperature and dew-point temperature at sea level on the leeward side of the mountain range (assuming all condensed moisture "rained out" of the air parcel during its ascent).

During lecture Steila would inform students that to prepare for this kind of examination, they should be able to think through certain steps to answer exam questions on the topic being discussed. He frequently demonstrated such a sequence of steps.

Some students, accustomed to multiple-choice tests, would panic when they first heard about Steila's exams. About 5 immediately dropped the course. The remainder learned that they could not compensate for a missed lecture (because their textbook provides facts but not a way of thinking about those facts); therefore, attendance improved with this method. Since there were few questions on his exams, Steila could easily grade 200+ papers. Although overall grades did not change, Steila felt confident his students retained more of an in-depth understanding of the material than if he'd used multiple-choice questions.

CHAPTER 6

COMMENTARY

Abigail Lipson, Ph.D., Senior Clinical Psychologist, Bureau of Study
Counsel, Harvard University

Early in the development of this project, I proposed a triadic model for under-standing the role of examinations in the dynamics of a course. The elements of the model include the "what" of a course (what students are expected to learn), the "how" of the course (how students are expected to learn it), and the "how well" of the course (the course's assessment/evaluation process). According to the model, the best educational outcomes are achieved when these three elements are well integrated with one another. Conversely, problems with a course can often be understood in terms of the lack of correspondence between the elements.

The focus of this project—examinations—nominally has to do with only the third ("how well") element of the triadic model. But it is clear from the participants' descriptions of innovative examination procedures in this and in Part 1 that they are strongly, if sometimes only implicitly, aware of how examinations operate in the larger context of course dynamics. These teachers do not conceive of examinations in isolation, but rather are concerned with the relationships between examination practices, course objectives, and the course's educational activities.

For example, Thomas Haas, formerly at the U.S. Coast Guard Academy (Part 1, p. 94), stresses the importance of making his chemistry course objectives and requirements explicitly clear to students at the beginning of each exam cycle. The objectives thus become a yardstick by which students can assess their own progress, determine what sort of help they might need when, and prepare adequately for classes and tests. In a very different context, Kurt Hollocher at Union College (Part 1, p. 137) incorporates the examination process seamlessly into the aims and activities of his mineralogy class: he hands out a box of 65 minerals to each student on the first day of class and the exam consists of the student's identifying all the minerals by the end of the semester, using the various assessment techniques they learn during the course.

These are just two examples among many in which an innovative examination owes its success not only to the design of the examination *per se*, but also to how well the examination corresponds to the objectives and activities of that course. In addition to this within-course consistency, what sort of consistency can be seen across courses? What do these innovative examinations in many different fields of study and course topics have in common with one another?

One common theme that emerges is that examinations should not only reflect the values of the course, but also should strongly correspond to the real-world activity of scientific inquiry. The innovations reported here and in Part 1 emphasize that science learning should resemble "real" science, that aspects of an apprenticeship model should be incorporated into science education, and that relevant skills should be learned by means of hands-on activities.

For example, Herbert Bernstein at Hampshire College (Part 2, p. 106) comments that his school's project-based examination system "puts research into the very beginning of each student's education." Fred Hartline at Christopher Newport University (Part 2, p. 112) designs exams composed of tasks that closely resemble real-world problem solving: "All involve basic principles that must be applied in novel situations using equipment with which they're familiar." June Oberdorfer at San Jose State University (Part 1, p. 141) administers a hydrogeology exam that provides students with field data and a referral question, so that the student's "end report is similar to a consultant's hydrogeologic investigation."

Instructors' efforts to bring together science-learning and science-doing raise another question: To the extent that their innovative examinations resemble the activity of science, exactly what are they communicating to students about the nature of scientific inquiry? The following lessons stand out:

Scientific inquiry is collaborative and interactive. Students are introduced to the collaborative and interactive nature of science through group work and communication, two common characteristics among the innovations. G. Marc Loudon at Purdue and Richard Bauer at Clemson (Part 2, p. 85), for example, have their students discuss chemistry exam problems with their study groups. Louden explains, "Students need to learn that science is a social endeavor requiring collaboration, and that real-world chemical problems are never multiple choice." Pascal de Caprariis of Indiana University (Part 1, p. 134) has students write short response papers which are graded not only on content but also on organization and grammar. When questioned as to the applicability of such standards in a science class, de Caprariis argues for the fundamental importance of clear communication—"life is an English class."

Scientific inquiry is active and self-motivated. Many exam procedures are designed to encourage students to take responsibility for their own meaning-making, rather than acting as passive recipients of "made" meaning. Harry Ungar of Cabrillo Community College (Part 2, p. 95), for example, requires of his students careful writing and rewriting of notes, and observes that his students "celebrate" the deep understanding they achieve through this process. Theresa Zielinski of Niagara University (Part 2, p. 152) involves students in study groups, structured note writing and rewriting, and collaborative blackboard work, and finds that "students are making a real effort to integrate the material themselves.... They seem to appreciate the challenge of a difficult question and to enjoy the sense of accomplishment in a well-crafted answer."

Scientific inquiry requires persistence. Competency-based tests that students retake until they pass, questions that are addressed more than once (e.g., in study groups and then in class), and "redemption" opportunities in examinations all serve to communicate to students that scientific inquiry is not a one-shot-only/right-or-wrong affair (which standardized multiple-choice tests seem to suggest), but rather a process of thinking and rethinking one's way through complex and sometimes indeterminate problems, visiting and revisiting material to make better and better sense of it. For example, Gene Lene of St. Mary's University (Part 1, p. 138) has his students complete a closed-book exam in green ink, then turn in the green pen for a red one and go over the exam again, correcting it while consulting their course notes. Lene comments, "In this way, students learn to recognize what they know as well as what they don't know, and the test becomes a far more valuable learning experience than those following a traditional one-time format." Valerie Keeling Olness at Augusta College (Part 1, p. 64) has students pair off to work out test problems, and then reshuffle into new pairs to consider the problems again. And Theresa Zielinski of Niagara University (Part 1, p. 125) has students write notes on the readings before coming to class and then amend these notes in a different color ink as they listen to the lecture.

Scientific inquiry involves reflection and self-evaluation. Many innovative exam practices encourage students to become more aware of the process of inquiry and better able to evaluate their own progress and performance. Gerald Van Hecke of Harvey Mudd College (Part 2, p. 97) gives students an open-book exam, requiring them to comment on their problem solving as they go about it. He instructs them to "indicate what you are doing as you proceed....While numerical responses are expected, the largest amount of credit will be given for the correct approach, formula, etc." Frederick Reif of Carnegie Mellon University (Part 1, p. 175) allows students to redeem missed test points by identifying their mistakes, diagnosing the likely reason for them, and offering corrections. Erica Harvey at Fairmont State (Part 1, p. 97) gives her students exam answers—complete problem-solutions or laboratory notebook entries—and asks her students to critique them, addressing specific mistakes or questionable assumptions.

Scientific inquiry entails, at all times, managing one's confusion and stress. Many instructors mention that their innovative practices serve to reduce students' anxieties, or help them manage them, while improving their performance. As a rule, performance anxiety is at its worst in situations in which the expectations are not clear, the stakes are high, and the outcomes are indeterminate (as with curved grades)—all conditions that are typical of traditional examination practices but are ameliorated by "innovative" practices such as the use of clear and explicit objectives, multiple formats/exposures, repeated opportunities for self-evaluation, and so on. Shirish Shah of the College of Notre Dame (Part 1, p. 119), for example, provides her students with clear performance criteria and awards absolute rather than curved grades. Sandra Palmer of Cazenovia College (Part 1, p. 66) gives untimed exams. And John Zwart of Dordt College (Part 1, p. 191) provides students with multiple opportunities for improving their performance and reports that students "shift from seeing me in an adversarial way; they recognize that I am more interested in their better understanding of the material than simply trying to get a good spread of grades."

These lessons about the nature of scientific inquiry are all lessons beyond the science (content) of the course, and learned only when instructors take care to teach them. DiLorenzo of Montclair State (Part 2, p. 41) acknowledges, "As instructors, we are responsible for the facilitation of the group process." Margaret Weck of St. Louis College of Pharmacology (Part 2, p. 60) and Paul Rab of Sinclair Community College (Part 1, p. 67) describe in detail the care they take to help their students learn how to work in groups. Similarly, Harry Ungar of Cabrillo Community College (Part 2, p. 95) and Theresa Zielinski of Niagara University (Part 1, p. 125) describe the efforts they expend to help students learn how to keep track of their understandings as they go through the semester.

One of the major questions raised by this project has to do with how widely applicable the innovations described here may be. Are they too idiosyncratic, too time-consuming, too cumbersome to be used by the average instructor in a large class? I suspect that, just as instructors want students to understand that real learning cannot happen in a *pro forma*, minimal-effort, passive way, administrators and instructors must understand that neither can real teaching. Just as the exam innovations described here provide students with reward structures that encourage real learning, academic institutions must make every effort to provide their faculty with reward structures that encourage real teaching.

APPENDIX

Sheila Tobias
Jaqueline Raphael
P.O. Box 43758
Tucson, Arizona 85733-43758

To: Friends and Colleagues
From: Sheila Tobias and Jacqueline Raphael, co-authors
Subject: Your Contribution to a Manual on Testing

The California State University System has recently inaugurated a unique university press, one that will serve the teaching faculty in post-secondary institutions with useful how-to manuals and handbooks. The first of these, *Computers in the Classroom*, was published in 1990 and has been distributed at near-cost to postsecondary educators in the California State University System and elsewhere. The second, now in preparation, will be a compendium and analysis of new ideas in testing practice for the teaching of college science entitled: *In-class Examinations: New Theory, New Practice for the Teaching and Assessment of College-level Science*

Why This Project?

It has been said that examinations are the "latent curriculum" that, more than what faculty *say* they want students to learn, drives student behavior. Yet, in most undergraduate science courses, the same testing practices have been used for years. This, despite much criticism. Reasons for not making change range from habit, the convenience of textbook-generated exams, the cost of hand-grading as against machine-grading, and so on. Still, the issue ought not be ignored. This manual will serve to put it on the teacher-innovator's agenda and serve to support that teacher–innovator when he or she approaches a dean or department chair for financial or other support.

What Is Planned

This 100-page manual will consist of three parts: Part 1, a 30-page analysis of in-class examinations as currently employed in college-level science courses; Part 2, a 60-page collection (and commentary) of innovative and exemplary testing practice at the undergraduate level in science (possibly mathematics); and Part 3, analysis, conclusion, and annotated bibliography, written materials, computer-related materials, contacts, and other resources. In addition, a list of funders with an interest in testing will be appended to the volume.

How You Can Become a Contributor

If you have designed and used any testing innovations, we would like you to contribute to this project. Do not, incidentally, let the word *innovation* intimidate you. We define it as a unique or experimental examination practice—in exam design, format, style, or grading—that enhances the quality of learning in your classroom. Here are some categories—by no means an exhaustive list—that have already been identified as areas for innovation:

1. Exam design: content of test items, test item construction
2. Exam format: verbal, pictorial, quantitative, open or close-ended, multiple choice, etc.
3. Exam "ecology": individual, group, in-class, take-home
4. Exam grading practices: pass/fail, curve/absolute/resurrection or other point- compensation schemes, "contracts"

Initially, we will simply be collecting contributors' names and a brief description of their innovation(s). Later, the authors will get back to contributors for more detail in a telephone interview and may ask for additional written commentary on how the innovation was developed, how it was received (by students and colleagues), whether it is continuing, and, if not, why not.

All Contributors Will Have the Opportunity to Edit and Approve Any Description of Their Innovations

You can help us by completing the enclosed brief summary sheet, which will give the authors all the information they need to start the collection and commentary process. There's room on that sheet for the names and addresses of other colleagues whose work deserves to be included in this book. If you need more information about this project, please feel free to call Sheila Tobias or Jacqueline Raphael at (602) 628–1105. Or, if you prefer, fax us at (602) 882–6973.
Please answer on this or separate piece of paper and return the summary sheets to

Sheila Tobias
P.O. Box 43758
Tucson, Arizona 85733-43758

Assessment Innovation

Name and Address (please include phone and fax)

Course(s) taught:

Describe what new examination form and/or practice you have been employing and for how long you've used it using the remainder of this page and the reverse side as needed.

Please list names and addresses of colleagues who might have something to contribute to this volume.